# THE ENLIGHTENING

Copyright @ 2017 Kevin Goczeski

All rights reserved. No part of this book may be reproduced in any form or by any electronic or mechanical means, including information storage and retrieval systems, without permission in writing from the publisher, except by reviewers, who may quote brief passages in a review.

# TABLE OF CONTENTS

The Enlightening .................................................................................................. 5
CHAPTER 2 ......................................................................................................... 8
CHAPTER 3 ........................................................................................................11
CHAPTER 4 ....................................................................................................... 13
CHAPTER 5 ....................................................................................................... 15
CHAPTER 6 ....................................................................................................... 19
CHAPTER 7 ....................................................................................................... 22
CHAPTER 8 ....................................................................................................... 31
CHAPTER 9 ....................................................................................................... 35
CHAPTER 10 ..................................................................................................... 41
CHAPTER 11 ..................................................................................................... 44
CHAPTER 12 ..................................................................................................... 48
CHAPTER 13 ..................................................................................................... 52
CHAPTER 14 ..................................................................................................... 58
CHAPTER 15 ..................................................................................................... 67
CHAPTER 16 ..................................................................................................... 73
CHAPTER 17 ..................................................................................................... 78
CHAPTER 18 ..................................................................................................... 88
CHAPTER 19 ..................................................................................................... 98
CHAPTER 20 ................................................................................................... 107
CHAPTER 21 ....................................................................................................112
CHAPTER 22 ................................................................................................... 122
CHAPTER 23 ................................................................................................... 132
CHAPTER 24 ................................................................................................... 138
CHAPTER 25 ................................................................................................... 149
CHAPTER 26 ................................................................................................... 160
CHAPTER 27 ................................................................................................... 171
CHAPTER 28 ................................................................................................... 178

| | |
|---|---|
| CHAPTER 29 | 185 |
| CHAPTER 30 | 189 |
| CHAPTER 31 | 200 |
| CHAPTER 32 | 209 |
| CHAPTER 33 | 215 |
| CHAPTER 34 | 221 |
| CHAPTER 35 | 228 |
| CHAPTER 36 | 232 |
| CHAPTER 37 | 245 |
| CHAPTER 38 | 248 |
| CHAPTER 39 | 253 |
| CHAPTER 40 | 257 |
| CHAPTER 41 | 266 |
| CHAPTER 42 | 278 |
| CHAPTER 43 | 284 |
| CHAPTER 44 | 292 |
| CHAPTER 45 | 299 |

# THE ENLIGHTENING

Prior to my Enlightening, I actually believed that scientists knew just about everything there is to know about the universe. It never even occurred to me that many great thinkers throughout history believed the same thing, quite mistakenly, in the context of their own times. From the brilliant scholars of the Enlightenment and Renaissance, to the most learned of the ancient Greeks and Egyptians, and probably all the way back to the astronomer-priests of Neolithic times, a presumption of near-total understanding of the cosmic firmament has been a recurring thread of human thought.

Ptolemy, the ancient Greek astronomer, was quite certain that the Earth was at the center of the universe. The planets, sun and moon all circled around the Earth, in perfectly circular orbits. Closest to the Earth in its orbit was the moon, with Mercury circling in a still farther out orbit. Following Mercury were, in turn, Venus, the Sun, Mars, Jupiter, and Saturn, with an infinite field of stars orbiting the Earth even farther away.

During the awakening of the Renaissance, Nicolaus Copernicus proposed the heliocentric universe, with a motionless Sun at the very center and orbited by the Earth, the moon, the other planets, and the stars of the firmament. Perfectly circular orbits were still the model, until the early 17th century, when Kepler came along and described elliptical orbits. At that point, there was not an inkling of the existence of galaxies, or that our sun was actually a star, or much more that we know today as common knowledge.

Today, we revel in our own intellectual certitude, while preferring not to focus on several disconcerting modern-day mysteries of cosmology. One mystery is that nobody on Earth can describe the physical mechanism behind the pervasive phenomenon of wave-particle duality. Wave-particle duality is undoubtedly a central principle of our universe, since photons of light, electrons, and even fairly sizeable molecules all appear to exist as both particles and waves, simultaneously.

In fact, it is not a big jump to wonder if every single thing in the universe exists as both a particle, and a wave, quite paradoxically. Even people and planets, perhaps, exist in this bizarre dual state. Yet even though the physical mechanism of this fundamental wave-particle duality has been a complete scientific mystery for more than 200 years, I still clung to my belief that we knew just about all there was to know about the universe!

Likewise, the paradoxical invariance of light speed, which is the Second Postulate of Albert Einstein's Special Relativity theory, had been enshrined as a scientific concept for over 100 years, with no apparent physical mechanism to back it up. I could add newer concepts such as dark matter and dark energy to the list of profound cosmological mysteries, yet still, in my own mind, our scientists knew exactly what was going on in the universe, for the most part. Quite a few physicists, if not the broad majority, would have agreed with that sentiment, I believe.

I suppose our human stubbornness drives us to frame things in terms of our own expectations, which certainly caused me to miss out on many great chances in life. If a new thought or idea was outside of those expected frames, I often rejected it, as a matter of course.

Quite simply, I had been overly content with my worldview, as must be true for most people. Who wants to question what they've known all their life? In the end, it took a series of extraordinary, eye-opening experiences, for me to contemplate any big changes in my own ways of thinking.

I am now completely convinced that there is much, much more to the universe than what is obviously visible or measurable. The unseen portions of the cosmos infinitely dwarf the portions that are seen, not only from our own human perspective, but from *any* single perspective. Truly, the physical foundations of our own discernable existence are laid down in a manner that neither physicist nor philosopher-- on this planet, at least— ever dreamed of before. The entire spectrum of what I experienced and what I learned, during those nine transformative days, is what I refer to as my "Enlightening".

# CHAPTER 2

After junior year of college, I was back in my home town of Rundle Heights for summer break. I worked part-time at the R.H. Enterprise restaurant, which was only a few blocks away from where I was staying, at my Cousin Walter's house. Walter was in Seattle for a few months, because of his job, and I was house-sitting, basically, while paying him a token rent out of my meager part-time paycheck.

Walter said it was the first time in his four years with the company that they had asked him to travel anywhere, and it just happened to coincide with most of my summer break, so the timing couldn't have been any better for me. Had I been back at my parent's house, in my old bedroom, as usual, the situation would have been a lot more complicated. As it was, my mom and dad usually stopped by three or four times a week, to bring me some lasagna or split pea soup or whatever else my mom made, and to make sure I was taking care of Walter's house, like I said I would.

The first question most people ask me about my parents is whether they had ever met Merle. Due to my parents' habit of showing up at my door unannounced, both of them did have interactions and conversations with Merle, on more than one occasion. I actually suspect that Merle purposely timed those particular visits, knowing full well that my parents were on their way over. At any rate, it was quite obvious that he enjoyed talking with them. I guess that's why my parents never had any suspicions about Merle, while he was around. They *liked* Merle.

Also, we lied to my parents, and told them Merle was a friend I knew from school.

That seemed to satisfy them enough to not ask too many questions. Merle and I didn't really enjoy it having to be that way, but there had to be some kind of cover story as to how I knew Merle, and being "friends from college" seemed like the perfect angle.

My mother says that Merle seemed totally normal to her, and that the whole thing was quite difficult to believe, at first. But I'm fairly sure she does believe it now, more or less. I don't think Mom's too clear on the physics aspect, though.

My father just completely steers away from the subject of Merle, as much as he can. He usually gets sort of angry when it comes up in conversation, and as far as I know, he has never even acknowledged the possibility of it all being true. So I try to not talk about Merle when my dad is around.

On the other hand, my dad *does* enjoy discussing the physics part of it with me, and most of the time, we manage to stick to talking about the workings of the universe.

Every once in a while, though, I slip up and accidentally mention Merle, and my dad will give me "the look" that all fathers have, I think, when you've overstepped some boundary or whatever.

As for me, I always thought that the idea of a hyper-dimensional universe stands on its own. Who really cares how I claim to have arrived at such a thing? I often wonder if anybody would have paid any attention at all, if Merle hadn't been included in the story, anyway. Merle himself told me that I could choose to mention him, or not mention him, after he left. He would leave that totally up to me, is what he said. He was just hoping I could figure out how to gain the attention of those with ears to hear, is how he put it. I promised him that I would do my best, and that was about how we left it.

That's why I decided to put it all down into words, as a documentation of the events that I experienced, and the ideas that I was exposed to, that summer. This is not only a chronicle of my own personal adventures and enlightening, but it also serves as my best "evidence" of the cumulative progression of reinterpretations necessary for an entire series of revolutionary new understandings in relativity, astrophysics, quantum mechanics, particle physics and cosmology.

# CHAPTER 3

I never wanted to pretend that Merle didn't exist, or that Merle didn't influence my Enlightening, or that Merle didn't place the theory right in my lap. I've been upfront about the whole thing all along, except for in the beginning, when I didn't want to mention Merle to my father, yet.

As I had mentioned, my father does embrace the physics itself, even though he doesn't like the Merle angle. My dad has a double major, in physics and engineering, so he studied relativity, quantum mechanics and all the rest. He was always especially interested in relativity, too, like I was.

I explained the theory-- or theories, I suppose-- to my father, after Merle left. I decided not to mention my unusual friend, at first, which was definitely a smart move on my part.

My dad came to understand the entire arc of this new way of thinking about the universe, quite clearly, as we talked about it. He tells me the thing about it is that you have to go through all the different areas of reinterpretation, one by one, with an open mind. Then, when you look back at it, you see that it all fits together, like a key in a lock, and you realize the universe suddenly makes a lot more sense than it did, before.

The way my dad puts it, I haven't killed, or even wounded, relativity theory. I just expanded it and made it stronger, as well as a great deal more understandable. I remember my dad saying to me, so many times, "I can under*stand* hyper relativity." By that, he meant that he never really could understand certain physical aspects of the universe, as described by special relativity, in particular. I do agree that hyper relativity is definitely much more understandable, because it clearly represents simple, fundamental truths of the hyper-dimensional universe; truths that are explainable in simple language, and which do not contradict each other.

That's why I'd be remiss if I didn't mention that I'm really just the *messenger* of the "theory". Even Merle is just a messenger. Nobody can own truths of the universe. They just are what they are, and either we know about them, or we don't.

My dad told me that he knew Hyper Relativity was correct when he heard my analogy—actually it was Merle's analogy—of the famous double-slit experiment. Without mentioning Merle, I described the unconventional experiment that Merle had physically performed for me. The rather humorous specifics of it made my dad burst out laughing, and some of the milk he was drinking came out of his nose. We both had a good long laugh over that, and my dad ended up knocking over the pepper grinder, by slapping the table, hard, as he was laughing. He laughed until he finally had a coughing fit. "It's so ludicrously simple!" he said, when he finally caught his breath. "I can see it all perfectly!"

I'll never forget how my dad stood in front of me that day, face to face, and about as bug-eyed as I'd ever seen him. He grabbed me by both shoulders, real excitedly, and he sort of shook me back and forth a bit, for emphasis, I guess. "You have to tell somebody about this!" he said. "You have to let people know about this, Ken! This is really important!" And that was before I even had a chance to tell him about all the rest of it, too.

# CHAPTER
# 4

Although I occasionally worked breakfast or lunch in the main restaurant area, I usually worked late shift at the Enterprise Lounge, so I was on a late-night schedule that summer, no question. That was part of what was so weird about the first day of my Enlightening.

It was a Wednesday morning. I woke up much earlier than usual that day, and I was actually feeling great. I had gone to bed quite early, the night before. I showered and got dressed and it was still only 6:15 a.m. The past few mornings prior, I'd probably slept in until almost noon.

I decided to take advantage of my uncharacteristically early start and take a nice bicycle ride. I bought the bike about 2 weeks before, and I was enthusiastic when I brought it home, but until that morning it had been sitting, unused, in the corner of the garage.

I would have still been in bed sleeping for probably another six hours, had I gone out with Ronny and Keith to the Cubs/Phillies game the night before, as I had planned on doing. I know they ended up going to a bar after the game, and they didn't get home until almost 3 a.m. I didn't feel well, though, beforehand. Mostly I just felt completely exhausted, and I could barely even get out of bed in the middle of the afternoon, which was bizarre. I was just too wiped out to make it to the game. So, knowing that Bryce was a huge Phillies fan, I let him go, instead of me. Passing up good seats to a ballgame was highly unusual on my part, to say the least.

That was the night that Ronny got hit in the arm by a foul ball, dropped his hot dog on his shirt, and spilled some beer on a girl named Ashley, all in the same pitch. I remember seeing the stitch marks from the ball, on his arm, more than a week afterward-- so you can imagine how hard the ball hit him. Still, he recovered in time to immediately chug the rest of his beer, which got a big laugh from the people in that section; or so I was told, anyway, by Ronny, who has been known to embellish just a bit.

That foul ball even made it onto TV, which I saw because I had recorded the game. The only thing they really showed on TV, though, was about six seconds of the big burly guy who ended up with the ball, after it had ricocheted off Ronny's arm. He held the mustard-streaked ball high above his head in one hand, and paraded down the aisle, in triumph, like he was on tour with Lord Stanley's Cup itself. With his other hand, he vigorously high-fived every single person that passed by. The TV missed the part, just seconds earlier, where he outwrestled an elderly man for the ball, and spewed the man's nachos all over the aisle in the process, according to my friends.

If you looked carefully, or paused the recording, you could find Ronny in the background, with a garish mustard stain on his sky blue shirt. He was about eight rows behind the burly guy, apologizing to Ashley for spilling beer on her. Eventually he got her cell phone number, too.

All in all, that had to be one of the more eventful foul balls I've ever heard of. Ronny ended up paying for the dry cleaning, and he and Ashley even went out on a few dates after that. On their third date, they went to a party, where Ronny met his future wife, Eva. She was a friend of Ashley's sister. Ronny always says that if that foul ball had gone a few inches one way or another, he'd probably still be single.

The Cubs ended up winning the game in eleven innings, and I know I missed out on a great time overall, but I'm not complaining. I wouldn't change what happened, for anything in the world-- or the universe, for that matter.

# CHAPTER
# 5

I didn't have a car of my own, at that time, or much money to speak of, either. My father paid for the new bicycle, which was much appreciated. There was a nice little lake just outside of town called Baxter Lake, and I'd wanted to ride out there, ever since I moved into Walter's house. So I was specifically thinking of riding out to the lake, when I got the bike. However, as I left the driveway early that Wednesday morning, I turned left instead of right, and headed off in the opposite direction. I didn't even think about it at the time. I just did it.

It seemed like I was going about my normal life, but it's obvious to me now that I was already being interacted with. I never specifically asked Merle about it, but I've always assumed that's how it happened. It doesn't bother me, either. I can understand why they did it. I sure don't doubt that they had the ability, either. They could probably do that as easily as I could sweep crumbs off a table.

It was relatively quiet at that early hour, and not many people were out yet. There were only a few cars and trucks on the road, and a siren was faintly wailing somewhere off in the distance. After riding on the road for a while, I came across a bike path, and I followed it. Eventually, the path curved away from the road and into a forested area, and the sounds of the street soon faded away completely.

Robins were chirping to each other, in their quirky, rhythmic way, across the grassy clearings at the forest edges. Grasshoppers darted all over the place, and squirrels crashed through the underbrush, as I rolled past. The muffled, soothing drone of a jet wafted and waned across the rustling leaves of the forest canopy, while the gears of my bicycle clicked and whirred, reassuringly, beneath me.

So it was nice, in an almost hypnotic sort of way. I was just thinking about how beautiful it all was, when suddenly it occurred to me that I had forgotten to head down to the lake. And I didn't care, at that point.

I continued cruising down the path, when it slowly dawned on me that I was hearing a new and different sound. It seemed to be wafting down towards me from out of the sky, or perhaps from the trees. It was a sort of droning, low buzzing sound, like a very deep musical note, almost. Although it seemed subtle, and took a while for me to notice, it was actually fairly loud, in retrospect. As soon as it entered into my consciousness that I was hearing the sound, it stopped completely. I momentarily wondered if it might be some sort of bird I had never heard before, like some type of large goose or something, flying overhead.

As I was pondering the situation, I rounded a bend in the trail and came to a spot where there was a decent-sized clearing on the left side. It was a pretty little meadow, maybe a few acres in size. Off on that side of the bike path was a water fountain, like they often have in forest preserves, and standing alongside the fountain were four people. Or rather it was one man standing out in front, looking towards me, and the other three people- women apparently- standing off to the other side, looking out at some deer that were across the meadow and near the forest's edge, not far from the river.

The man was maybe 6'3", with bright blonde hair. He was dressed in a "Miami Heat" t-shirt, red basketball pants, and orange and black basketball shoes. It looked like he was ready for a game, or a run, or something.

The three women were all wearing white hoodies, which seemed unusual, right off the bat. It was fairly warm out, even at that early hour, but they still all had their hoods up. When they turned towards me, I could see that one woman was a blonde, one a redhead, and the other had black hair. I remember thinking that they must all be into sports, because each hoodie had a different sports logo on the front; one was an "L.A. Galaxy", one was a "New York Giants", and one was a "Dallas Stars", I recall. They all had similar whitish jogging-type pants, which looked like simple cotton pants, with maybe some pin striping in black or grey. I noticed that they all had very similar black and white athletic-type shoes, also.

All in all, their outfits were so similar that I thought they might have been teammates on some kind of sports team, or something. And maybe the tall man was their coach, or something, since it seemed like all four were probably together. I noticed that one of the women was holding a wooden hiking staff, which was a little weird, I thought.

Right then, as I approached the group, my bike started getting very hard to pedal, like it was freezing up on me. I had to hop off my bike, maybe 20 yards from the group, because I couldn't get the pedals to even move at all.

At that point, the man in the Miami Heat shirt totally surprised me when he called out to me. "Excuse me! Mr. Sylvanewski!"

Now, as far as I could tell, I had never seen this guy before in my life. So I found it pretty startling when he sort of came out of nowhere and called my name. At first I didn't say anything, and I just stood there, straddling my bike.

"Mr Kenneth Sylvanewski!" the man said, and took a few steps towards me.

Suddenly I realized I was sort of vulnerable, out alone in the forest with a bike I can't pedal, and four people who could be ready to roll me, or whatever. I pictured myself getting pummeled by the lady with the wooden staff. And I felt especially uneasy about the one with the black hair. But there was something about the man that was non- threatening, right off the bat. His friends were now paying attention to me, but they were still standing off in the background, observing. If they were going to roll me, they were taking their time about it.

So I just said, "Yes", and hopped off my bike, as he continued to approach. Most people call me "Ken", or "Kenny", so I figured whoever this guy was, he didn't know me very well.

"We are friends of the man whose life you saved, from the car."

Well, that was a big surprise! I had been fairly successfully trying to put that whole episode out of my mind, lately, and now this guy I'd never even seen before comes up out of nowhere, and mentions it.

# CHAPTER 6

It was about six and a half years prior, in December of my freshman year of high school. I was walking along a sidewalk in the early evening, just outside the downtown area of Rundle Heights, when a car jumped the curb behind me. Actually, the driver had accelerated, and not used the brake, as the police found out when they looked at the car's computer, afterwards. The driver claimed that he hit the brake, not the accelerator, but the police thought that he may have briefly passed out from some prescription medication he had taken before he left his condo that evening. The whole thing was a horrible mess, really.

I heard the car thump as it crossed over the curb. When I turned to look, it was quickly bearing down on me, along with a few other people also on the sidewalk. I shouted out a warning to the others and dove out of the way, back towards the street, as the car, now completely up on the sidewalk, hurtled past. As I dove, there was another man right there, walking along the sidewalk and also in the path of the car, so I instinctively grabbed him and pulled him along with me. We both tumbled into the gutter area of the street and ended up in a cold, salty puddle. The car must have missed us by less than a foot.

The car did hit two people, and it continued on, taking out half the front stairs and awning of an apartment building. One young lady didn't make it, and passed away before she even made it to the hospital. I got lucky. And so did the guy I grabbed, because he definitely would have gotten smacked, too.

I stuck around for at least an hour, afterwards. It was a gruesome scene, with blood on the sidewalk, and the nightmarish, flashing strobe lights of squad cars and ambulances harshly illuminating the scene. The driver of the car was crying and wailing, hysterically, as the police put handcuffs on him. I spent some time talking to the detective about exactly what I saw, and what I thought happened. I think I was half in shock the entire time, myself.

My new acquaintance, who I had just tackled into the gutter, answered some questions from the detective, as well. I remember that the man thanked me, profusely, and told me how much he appreciated my "selfless behavior". I told him anybody would have done the same thing, and I apologized for getting his nice camel hair coat wet. He said that he didn't care about the coat getting a little wet, and that he was just glad to be alive.

The strange thing was that, just as I was wondering what was up, exactly, with this guy in the camel hair coat, he seemed to just disappear into the night air. One of the investigators wanted to ask him a few more questions, and he was gone. Nobody had seen him leave. As it turned out, he apparently gave the first police officer a fake name. I thought I might see him around the neighborhood sometime, but I never saw him again.

A couple months afterward, one of the police officers told me I probably would have gotten some sort of public Citizen Award for saving this guy's life, but since he gave a false name and nobody knew who he really was, that part of the story would have been a little embarrassing for the department. As it was, there were some articles in a few newspapers, and I got a nice letter from the Police Chief, personally thanking me.

But otherwise, my rescue of the mysterious man in the camel hair coat became just a funny joke for my friends. They had their own theories about it, like maybe I saved some sort of international terrorist, who was on the lam with a fake identity, and the driver was really a secret agent who was trying to take him out, and I ruined the whole thing. They made up all kinds of crackpot theories like that.

It's funny that, as it turned out, the actual truth was a whole lot stranger than anything my friends came up with. To this day, I wouldn't doubt that more of my friends believe the made up story about the international terrorist, than believe the real story.

Anyhow, I hadn't heard anything about the accident for quite some time. So I was pretty shocked when this guy by the fountain called out my name and then mentioned the man I had rescued that night.

# CHAPTER
# 7

By now, he was standing next to me, holding out a gloved hand to shake. He was wearing gloves that looked like they might have been riding gloves, although the group appeared to be on foot. Eventually, I found out that the gloves were specially designed to collect biological samples from my hands or clothing. Also they helped protect Merle, I guess, who probably had to be careful about what he might pick up from somebody like me. Merle had spent the equivalent of several years preparing for the trip, which included being inoculated against whatever strange brews he might come across in his adventure. He had already spent a lot of time on-planet, but he was still being careful. So we shook hands, the man with the Miami Heat shirt, wearing gloves, and me barehanded.

"Mr Syl-"

"You can call me Ken."

"O.K.,'Ken', then." He smiled at me, quite sincerely. "We are here to set up a cross- cultural exchange. Our entire… organization… is very grateful for your actions in saving our friend."

"Who was your friend, exactly? He gave the cops a fake name and then he took off!"

"We came from a great distance. Our friend, Magu" (I later found out that his full name is Magu Yahay, or something to that effect), "was setting the groundwork for our trip. He was distracted, and your attention was very sharp. You heard a noise, and you reacted very quickly. Magu was a little lost in the wilderness, I'm afraid, and did not react." Merle looked back at the three ladies, who by now had strolled up behind him.

All three of them sort of smirked or smiled at Merle, like there was some sort of inside joke going on.

The tall blonde-haired man continued speaking. "As I was saying, we are setting up a cross-cultural exchange. We have been searching for a likely participant from your side. Your actions in saving our friend attracted much interest." He smiled again. "And finding out that you are a rather open minded student of Physics was the--" he seemed to search a bit for the words-- "the icing on the cake."

Well, as soon as he said that, I figured that this was some kind of joke. First he mentions the runaway car incident, which was strange. And then he mentions me being a Physics major, which people always seem to find humorous, for some reason.

"Where exactly did you say you guys are from?" I asked the group as a whole. I was already getting tired of the gag or scam I thought they were trying to pull on me, and I was quickly developing a bit of an attitude about it.

"I'm not sure I want to say, just yet. But from very far away." I laughed. "Probably outer space, right?"

He didn't laugh at my joke. Playing it up well, I figured. I wondered who set them up to this. Probably Bryce or Keith, or maybe even Ronny or Cam or somebody else from school, I guessed.

"Would that disturb you?" he asked. "If we were from outer space?"

I laughed even louder. "No, that wouldn't bother me at all. If you're some kind of alien, maybe you'd be willing to tell me the secret of the universe or something."

At this, the three ladies started whispering back and forth, and the blonde pulled something out of her hoodie pocket. I took a step back, just in case it was a gun or something. I was surprised to see that it appeared to be a plastic bulb with buttons or something on it, about the size and shape of a hand-grenade, and with a spindle, maybe three inches long, coming out of the top. For a few alarming moments, I was actually considering that the thing *might be* some sort of electronic hand grenade or something.

The blonde woman held it out in front of herself, in her right hand, and then the bulb-- or the top of the bulb, with the spindle, at least-- started rotating slowly and then spinning really fast. With the fingers of her left hand, she tapped some buttons or something on the lower portion, while the other three watched. I had no idea what was going on. I thought it must be some kind of toy or weird phone or something, but I couldn't really figure it out. Suddenly it stopped spinning, and the spindle swiftly retreated back into the body of the bulb. The way it went back in so quickly and smoothly freaked me out, and I took another step back. Then she quickly slipped the bulb back into her pocket. I looked up, then, to see that all four of them now looking at *me*. By then I was feeling quite uncomfortable.

The tall blonde-haired man spoke again. "Excuse me for not properly introducing ourselves." He told me his name was "Mer-ell", or something that sounded a lot like "Merle" to my ears.

"Merle, did you say?" I asked. Maybe he was ready to give me some clues as to who was putting him up to this.

"Sure", he said. "'Merle, that's *good*." He seemed weirdly satisfied with the name, like I was naming him Merle for the first time or something, even though he himself said that his name was Merle. Or something close to it, I guess. "And these are my associates", he said, waving his arm in the direction of his three women companions.

Their names were all extremely odd. At the time I thought it was all part of the joke. "Clotro" was the one who did the weird stuff with the bulb. Actually, now that I got a better look at her, I realized that she was actually quite beautiful. She looked like a young Scandinavian fashion model or something, with long blonde hair streaming out from under her hoodie, bright blue eyes squinting a little bit in the early morning sunshine, as she peered at me. She smiled at me and said "hello" very quietly, in an accent I couldn't quite place, although I was thinking maybe Swedish or something like that. Even though I wasn't too happy, at this point, I couldn't help smiling back, just a bit.

Then there was "Latsis", who held the walking staff, which appeared to be a natural wooden hiking staff; just a straight piece of polished, knotty wood. She seemed to be maybe in her late thirties to early forties or so. I figured she was probably big into hiking, but it was odd these days to see the wooden staff, instead of a modern hiking pole, or set of poles. She certainly looked like she was in great shape; as they all did, in fact. Latsis tipped her head in a slight bow and smiled. She had bright red hair, maybe shoulder length or so, and she had a dark olive complexion and amazingly green eyes that gave her a stunning, exotic look.

The third woman was "Atropha". She appeared to be the oldest of the three, with a bit of a weathered face like a sailor or something might have, but still she was not unattractive. She had jet black hair, cut short or maybe bunched up underneath her hood—I never could tell. Her eyes were so large and deep and black and serious, you felt like if you weren't careful, you could just tumble inside, and you'd have a heck of a time climbing back out. And I mean, you'd be frantically trying to claw your way back out of there. When our eyes met that first day, I wasn't able to look at her for more than a few moments, without feeling uneasy.

I believe that Atropha enjoyed her ability to make other people feel uncomfortable. I was definitely afraid of her at first, but now I think that she was just trying to help get me out of my comfortable place, and comfortable ways of thinking. Atropha didn't smile, but just tipped her head in a faint, almost imperceptible nod in my direction.

Their names were strange, but I assumed that "Merle" was just making up gibberish- sounding names, trying to make them sound alien or foreign, to make their scam or joke seem more realistic.

"Nice to meet you," I said, without meaning it very much. I nodded to them as a group.

Merle continued. "I cannot give you the secret of the universe. The universe does not hold any secrets. But it's interesting that you mention that."

Here's the punch line, I thought. Keith or Ronny were probably hiding somewhere and filming the whole thing to post online and make fun of me. It hadn't occurred to me at that point that Keith or Ronny must have been fast asleep in their beds after the eleven inning game the night before, plus the late night visit to the bar. The odds of them being awake at this hour, trying to prank me, would have been essentially zero.

"Are you familiar with Mark Twain?" Merle asked.

Well, I certainly hadn't expected a Mark Twain reference coming out of left field. "Duh, of course I know who Mark Twain was."

"He once said that education consists mainly of what we have *unlearned*. There is a lot of truth to that."

"So what's your point?" I was quickly losing all my patience with the entire charade.

"Well, let me ask you. Do you think that you understand the Lorentz transformations?"

The Lorentz transformations-- three separate yet related equations of astrophysics-- were originally formulated by the great Dutch physicist H.A. Lorentz, and they were cornerstones of Albert Einstein's Special Relativity theory. I like to think about the equations in terms of an observer-- a man standing on the earth's surface—and a traveler, who has blasted off from the earth and has accelerated into relativistic velocities. Relativistic velocities are velocities that are approaching the "speed of light", which is 300,000 kilometers per second, or 186,000 miles per second, and is commonly shortened to "$c$" in scientific parlance.

We spent many hours, over the next week or so, discussing relativity theory, and particularly these three equations, the Lorentz transformation equations. It took me a while to understand what Merle was getting at, and eventually I came to understand his central point.

The length contraction transformation indicates that the relativistic traveler will become shorter in appearance, from the observer's perspective, as he approaches light speed, or $c$. The mass increase transformation indicates that the traveler will seem to become increasingly massive, as he approaches $c$, from the observer's perspective.

And the time dilation transformation indicates that less time will pass on the traveler's clock, than will pass on the observer's clock, as the traveler accelerates to relativistic velocities.

Merle explained to me that both Lorentz and Einstein, and basically every other physicist on Earth, considered the transformations to be a complete description of the full range of possible acceleration in the universe; meaning that the equations were considered to describe the entire universe. It was considered impossible for a traveler to accelerate beyond the speed of light, since that would give the traveler infinite mass, compared to the observer. And it was also considered impossible to travel beyond the speed of light, since the length of the traveler would then become zero, compared to the observer. And it was also considered impossible to travel beyond the speed of light, since the passage of time between the two would literally have no meaning, or comparison, at that point.

In time, I have come to understand that, contrary to Einstein's interpretation, the transformation equations do *not* describe the universe in its entirety. They do, however, describe the full extent of the four dimensional frame of space/time—the perceptual frame of reference-- as experienced by any single observer, at whatever velocity that observer may be traveling, relative to any other object in the universe. That is, if a person or an object is traveling at velocities below $c$, relative to the observer, the length can still be measured, mass can be calculated, and the passage of time can be compared. The observer is able to perceive and physically interact with the relativistic traveler.

Albert Einstein developed an equation, known as the relativistic addition of velocities, to deal with the obvious problem presented by his description of a universe that was so oddly limited, in terms of velocity. The problem was that of a traveler with a jet pack, who accelerates to a velocity close to the speed of light, and who has another, smaller traveler stashed in his backpack. The smaller man also has a jet pack, and after the first traveler gets up to speed, the smaller man crawls out of the backpack and takes off, at a velocity close to the speed of light, compared to the first traveler. One can imagine an entire series of smaller men with jet packs, each traveling at nearly the speed of light, compared to the one before, which would be perfectly allowable, even with Einstein's interpretation of the transformations.

The problem was that the apparent total velocity of the second traveler, relative to a "stationary" observer back on Earth, would obviously exceed the speed of light, according to our Galilean way of looking at the world. In fact, it would be nearly twice $c$. That apparent contradiction to the "maximum speed of the universe" necessitated a "fix", which was the relativistic addition of velocities equation. That equation, rationalized by Einstein to maintain a single-reference frame universe where $c$ is the maximum velocity, falsely (and counter-intuitively) demonstrates, "mathematically", that the second traveler, from out of the backpack, is still traveling at less than the speed of light, relative to the observer. The same equation, in fact, indicates that all of the ever- smaller men with backpacks would still be traveling at less than $c$, relative to the original observer back on Earth. Merle pointed out, early on, that this was one aspect of relativity theory that had never been experimentally confirmed. In fact, Merle told me, several times, that "the relativistic addition of velocities was Einstein's greatest error." Merle once said that you could probably formulate an equation that shows that the moon is made of cheese, but that doesn't mean it's true. In other words, the fact that you have a mathematical equation doesn't necessarily mean that it describes a real phenomenon.

In more recent times, astronomers have found that in any direction you look, the farthest galaxies are receding from Earth at velocities just shy of $c$. So if you looked to the left, at a distant galaxy traveling at nearly $c$, and then looked to the right at a distant galaxy traveling at nearly $c$ in the opposite direction, you might assume that the two galaxies are easily traveling at velocities greater than $c$, relative to each other. Due to the relativistic addition of velocities equation, however, astronomers have been able to rationalize that the two galaxies are not "violating" the Lorentz transformations by traveling at velocities above $c$, relative to each other.

The reality, however, is that those two galaxies exist in completely different space/time reference frames, relative to each other. An observer in either galaxy would be able to see our Milky Way galaxy at the end of their range of vision, but neither would be able to see the galaxy on the opposite side, due to the velocity differentials between these two galaxies that are, in actuality, nearly $2c$, contrary to what the relativistic addition of velocities would indicate.

Even though I had always thought that the relativistic addition of velocities was very strange, I still found Merle's denial of the equation to be extremely difficult to believe.

But the more we talked about it, the more I began to see the truth in Merle's assertions. "That equation really doesn't affect any other mathematical equation in the entire theory of Special Relativity," Merle said. "So if we remove it, nothing is really affected, as far as the balance of the original theory goes.

"Also, in the hyper-dimensional universe, it makes sense that a traveler has an infinite mass, relative to the observer, when the traveler travels faster than $c$." I was obviously skeptical, and Merle got a little deeper into it. "What does it mean," he asked me, "to say that something has mass?"

I shook my head. "I don't really know, exactly." I think I probably could have come up with a definition, but I was curious as to what Merle was getting at, exactly.

"Think of it this way," Merle said. "The more massive an object, or a traveler, is, the harder it is for an observer to move it, or him. Something that is not very massive is easy to move. Something that is very massive is difficult to move."

"OK."

"Once the traveler accelerates close to $c$, he gets very difficult to move, from the observer's standpoint, due to being so massive, as indicated by the mass increase equation. The observer would probably need some highly advanced and very powerful equipment, just to alter the course of the traveler a tiny bit, if the traveler is approaching the $c$ velocity."

"OK." I don't think I truly understand the point, really, so much as I knew it made sense in light of the equation.

"But once the traveler exceeds $c$, he is no longer in the same four-dimensional frame of space-time as the observer. Which means the observer can no longer physically interact with the traveler, at all. So if the observer can't even touch the traveler, he certainly can't move the traveler- not even the least tiny bit, even if he has the most powerful equipment in the universe."

"OK."

"That means that the extra-dimensional traveler, who has accelerated to velocities beyond $c$, is infinitely massive relative to the observer. He can't be moved at all by the observer; not the slightest bit. That is the very definition of an infinite mass, *relative to the observer*, of course. In reality, though, in the traveler's own space/time reference frame, the mass of the traveler has never changed."

I said "OK" to that, also, but it took many conversations rehashing the same basic information before I truly understood that Merle was correct, and that it did make sense that a hyper-dimensional traveler would represent an infinite mass to an observer. Not that an infinite mass, as "perceived" by the observer, would cause any real problems. There is no way, in the universe, to directly interact with a hyper-dimensional traveler, other than, in certain cases, by rotating fast enough, artificially. That is just one of ways in which the universe works so beautifully. I have come to see that if there is anything in the universe that makes sense, it is the universe itself.

# CHAPTER
# 8

That first day of my Enlightening, I was obstinate and did not genuinely try to understand what Merle was saying about the transformations. Merle could see that I was not even making the attempt.

"Do you at least understand," Merle asked me, "that those equations involve the changing physical perceptions between an observer and a traveler, when the traveler accelerates to relativistic velocities?"

"Sure," I shrugged.

"Then answer me this." Merle rather slowly and dramatically reached his left hand out, and folded it back to his chin in a pensive gesture of deep thought. "If the Lorentz transformations have such a primacy to the theory, then why are some of the tenets of Special Relativity set up as universal tenets? Why is the single speed limit of $c$ set up as a universal tenet, according to your Special Relativity theory, when the transformations are only about the physical perceptions between two observers? And is it a reasonable jump of logic to simply make light speed inviolable for all observers? Or might there something more to it?"

The Second Postulate of Special Relativity states that the speed of light is always the same-- $c$-- regardless of the motion of the light source or the motion of the observer. That, to me, always did seem impossible to wrap my head around, in terms of it being a logical concept. After all, if I am in a golf cart traveling at 10 miles per hour, and I toss a ball at 15 miles per hour in the cart's direction of travel, a bystander on the street would measure the speed of the ball at 25 miles per hour, relative to him (the total combined velocities of the ball and the golf cart), while I would measure it at 15 miles per hour, relative to me in the cart. But light doesn't work that way. It always manifests itself at the same velocity—$c$—to all observers, regardless of their circumstances. I always had figured that it only seemed strange because we are used to seeing the world the way Galileo did, in terms of classical mechanics. I eventually came to think that there really was no understandable logic to it, and all you could say was "that's just the way the universe is." Other than that, there just isn't any way that anybody—Albert Einstein included, I'm quite sure—has ever truly understood what mechanism of the universe could possibly drive the Second Postulate.

Merle seemed to stroke an imaginary beard just a bit. "That part of it doesn't seem to make much sense, does it?"

"Oh for gosh sakes. Come on, man." Now I was really exasperated with the conversation and the ongoing gag. "You can give it up now." I started looking around for Ronny, or Keith, or Bryce, or anybody else I might recognize, somewhere nearby. But we were still alone in the meadow, at least as far as other people went.

Merle just continued looking at me quizzically, while Latsis leaned over and started whispering to the other two women.

"Ken," he said, while putting his other gloved hand on my shoulder, and in the process gathering more samples from my clothing, "this is not a joke. We are here on very serious business." I sort of shoved Merle's hand off my shoulder and twisted away from him, but he just continued talking. "It is not a joke, but nor is it an open-ended gift, either. There is a great reward possible, greater than you might imagine, but you will have to put some effort into achieving it. We can help you, but you--" again he searched for the phrase-- "you have to meet us half way. So please give some thought to what I have said about the Lorentz transformations and Special Relativity."

Now I was thinking that maybe somebody from my Physics classes was playing the joke. Because freshman year of college,

I had said in class that the universal speed limit of $c$ seemed hard to believe, on the surface of it. Because what can make you stop or slow down if you want to keep going faster? Our professor, Professor Thomas, shot that comment down pretty quickly by pointing out that any velocity greater than $c$ would violate the terms of the Lorentz transformations. Then another time I said that it seemed impossible that the vast entirety of the universe-- untold quintillions of tons of matter, and other forms of mass-energy in vast galactic scales-- was born out of a tiny little cosmic egg, apparently, that exploded in the Big Bang. He just laughed at that, like it was goofy to even question the idea. After that, every time he introduced a new subject, for the rest of the year, he'd poke a little fun my way. "Any objections to that, Mr. Sylavnewski?" he would ask. After 20 times, it wasn't very funny anymore, but he still got a laugh from the class every time. It did teach me that nothing good can come from questioning dogma. Now, however, I know that questioning dogma can be the first step on the path to an Enlightening. "Where did you guys say you were from?" I asked Merle.

Merle smiled at that question and pointed up at the sky. "Oh come on." I wasn't buying it.

"No," said Merle. "Look up there. Right *there*." And he pointed again, to a specific area of the sky. "Look up."

I looked up, and there was actually something up there, far up in the sky. It looked like a small, triangular black kite. It must have been high above the nearby clouds, but at the time I guess I didn't really think about that.

"And what is that?" I asked sarcastically. "Is that your spaceship?"

"As a matter of fact, yes, it is."

"Come on, man," I told him, shaking my head. "That's just some kid's kite." At that point, I really felt like I'd had enough nonsense. "I gotta go, anyway."

"We will talk again, very soon," Merle said. "Please give some consideration to what I am saying about the Lorentz transformations and Special Relativity. Take a second look at the Second Postulate and tell me if you think it really makes sense. And ask yourself how or why the universe would limit your velocity at all!" He looked at me again, very fixedly and serious.

"Oh, sure", I said, as sarcastically as I possibly could. "I'll be thinking about it, all the time." I reached back for my bike and turned it around to leave. Just then I realized that I couldn't move the pedals, anyhow. They were still jammed.

I looked back to see if the four weirdos were watching me, but they were already walking away, out into the meadow. I bent down to look at my bike to see what was jamming it up. I figured maybe a stick got caught in there or something, but I couldn't see anything out of the ordinary. Then I snuck a peak back at the meadow, and it was the damnedest thing. The weirdos were gone; completely and absolutely vanished. I looked all around, and they were just gone. They must have high-tailed it out of there, I thought. But how could they run fast enough? It was the strangest thing.

Just then, I realized that my bike was rolling free again. That was weird, also. I guessed the stick must have popped out on its own. So I hopped on my bike and headed back towards the house. I had enough of the bike trail by then, anyhow. In fact, I told myself I'd never come this way, down that bike trail, again. No way.

Just as I was headed back around the turn on my way back home, a young couple on their bikes came around from the other direction and passed by me going the other way. I thought I heard the girl say something like, "I wonder what was wrong with our bikes?" as they passed by, but I didn't really pay much attention at the time. I was thinking about my own strange morning.

# CHAPTER 9

After I got back to the house, I tried to figure out who was playing the joke on me. I texted a couple of people, but it was way too early. Keith didn't wake up until 12:30 p.m. and Ronny didn't answer my text until almost 2 o'clock. They both denied knowing anything about it. I talked with a few people from school, and they also denied it. They all seemed to not have any idea what I was talking about. In fact, when I mentioned the thing about the Lorentz transformations and the Second Postulate of Special Relativity, I think they thought I was trying to play a joke on *them*. I got the impression that my friends from school thought I might be a nut, getting in touch during the summer and telling some crazy story involving Special Relativity. So that sort of dissuaded me from trying much harder to find the joker who was setting me up. "Setting you up for what?" my friend Cam, also a Physics student, asked me. That was a good question. In telling the story, it didn't seem like a very funny joke. It was just plain weird, but nothing more than that.

After a while, I gave up my search because I was getting nowhere. Plus, I had to work that evening, and I was sort of tired, so I took a nap before I went to work, which I had never done before. I worked until past midnight that night, and didn't get to sleep until about 1:30 a.m. Yet I woke up early again the next morning, again feeling good. I made some coffee and had a bite to eat and got ready to take another ride, like it was the natural thing for me to do, which it absolutely wasn't, of course. I should have slept in until late morning, if not noon, according to my usual schedule. Instead, I was going to take the ride down by the lake that I had intended to take the previous morning. But you can probably guess which way I turned, instead.

I don't even remember the trip over there. All of a sudden I was enjoying listening to the creatures of the forest, and making the turn by the meadow. I started humming to myself, and it dawned on me that I was subconsciously echoing that same droning buzz I had heard the day before. As soon as I realized that I was hearing the sound again, I stopped humming, and the droning buzz stopped, also. As I wondered if I was just imagining the sound, I saw my four strange acquaintances, standing by the fountain and obviously waiting for my arrival. Again, the three women had white hoodies on, and Merle looked like he was ready to go for a jog, or to play basketball or something.

Merle waved a hand to me. "Hello Ken!"

"Hello, Merle." At that point I sort of realized where I was again. I was confused, and maybe a little scared, about how I got there. It was even earlier than the previous morning's "meeting."

"Ken, have you given any thought to the Lorentz transformations? The Second Postulate?"

"Ha ha ha. As a matter of fact, Merle, no. I did not."

Merle looked at me with a mixture of befuddlement and sadness, it seemed. He turned and held a hand up to Atropha, who was glaring at me. "Nobody said this was going to be easy," Merle said to Atropha, even though she hadn't said a thing. Then he turned back to me.

"Ken. Let's talk about those transformations. Just remember, the transformations are all about the perceptions of the observer, Ken. That doesn't mean that the mass or length of the traveler has changed at all, in any real sense. The passage of time itself is perceptual, in a sense. It's all just relativistic perceptual strangeness, between the observer and the traveler."

"But time passes more slowly for you," I protested, despite my ambivalence. "When you come back down, you've aged less than I have. Less time has passed on your clock."

"Well, that is true. That is very correct." I could see that Merle was pleased that I was at least thinking about special relativity. "Because the perception of time has permanent physical ramifications, in a way that length contraction and mass increase do not. Time is relative, but time is also an arrow, as they say. It's an arrow in both directions, actually, but the passage of time is permanent and real. Length and mass are just relative, in both directions. Although they both can get a little complicated, as well."

"I really need to get going," I said. I was feeling uncomfortable with the way the conversation was going. A bunch of double talk and nonsense, it seemed. But Merle wasn't stopping. He was persistent, I can tell you that.

He discussed why the length of the extra-dimensional traveler would be zero, or non- existent, according to the perceptions of the observer. After the traveler leaves the perceptual space/time frame of the observer, by exceeding the velocity of $c$ relative to the observer, he becomes not only untouchable, but also completely invisible, to the observer. "If the traveler is invisible to the observer, does that not describe a length of zero?" he asked me.

At that time, I still was not wrapping my head around these concepts, but as we discussed the matter over and over again, in ensuing conversations, I came to understood that, again, what Merle said made total sense.

Merle also discussed the relativistic weirdness of time perception. He pointed out that as a traveler reached $c$, relative to the observer, time would literally be standing still for the traveler, relative to the observer. "Which makes sense," he said. "At light speed-- as you call it-- the traveler is matching the outgoing time frame of the observer precisely, which would make the traveler's passage of time appear to stand still, from the observer's standpoint, until the traveler slowed down or turned to come back. And as the traveler exceeds $c$, there is absolutely no way for the observer to directly compare clocks. That makes sense, because just as the observer can't see or touch the traveler, nor can he measure his passage of time."

In ensuing conversations, Merle explained to me how the passage of time is measured by extra-dimensional space travelers, and how extra-dimensional communications are achieved. Basically, you always need links in the chain. Each link is another ship, or a transmission relay, that is traveling at less than $c$, relative to the traveler. Innumerable transmission relays are scattered throughout the universe by the various space-faring societies, traveling at vast ranges of velocities. So if a traveler is traveling at one and a half times $c$, relative to an observer, all the traveler has to do is link up with a relay that is traveling at an intermediate velocity, for example three-quarters $c$, relative to the observer. The relay, which is traveling at less than $c$ relative to both the traveler and observer, can indirectly link the two and transmit data between them.

If the relay rotates at nearly $c$, as the great majority of the probes do, according to Merle, the range and speed of data transfer is expanded even more. A relay spinning at nearly $c$, and moving away from the observer at nearly $2c$ (twice "light speed"), can still link up with the observer, since one side of the relay is spinning back towards the observer at nearly $c$. That means that although the probe itself is traveling away from the observer at nearly $2c$, a portion of the probe—that portion spinning back towards the observer-- is traveling at less than $c$, relative to the observer. By the same token, a portion of the relay is traveling nearly $3c$ (three times the speed of light), relative to the observer, and that portion can link up with a traveler who is traveling at nearly $4c$ (four times the speed of light), relative to the observer. So the relay can receive information from the portion rotating in one direction, and transmit information from the portion rotating in the opposite direction, at a greatly increased velocity compared to the original transmission. Without a relay of some sort, traveling and/or rotating at an intermediate velocity, an observer would have no idea where an inter-dimensional traveler had gone.

On that day, the second day of my Enlightening, I wasn't really quite ready to discuss or embrace any of these concepts. I was getting more and more uncomfortable, in fact, and I finally spun around to leave.

I was stopped dead in my tracks by Atropha, who had snuck in close behind me, without my realizing it. She stood there, mere inches from me, physically blocking my path to the bike.

Speaking of me feeling uncomfortable, that is when I discovered Atropha's angry voice. It was a piercing shriek and a thunderous roar at the same time, I can vouch for that. It seemed to hit you right in the solar plexus, and reverberate throughout your spine and out through the top of your skull. Atropha stepped up and got right in my face. She gave me a little advice, the hard way.

"*You should listen to him! He is trying to help you!*" She waggled her index finger at me as I took a few steps back to clear some personal space. She crossed her arms back across her chest, and it was dead silent in the meadow and surrounding forest.

No birds, no crickets, no chipmunks; nothing. Atropha stood in front of me with her feet spaced widely apart, and stared me down.

"This man," she said more quietly, waving her arm in Merle's direction and obviously trying to maintain her composure, "has *chosen you*. He has spent many years of his life learning your ways. He has traveled *97 light years* to come here and *help you*! And you want to just take off on your bicycle every time he asks you to *think*!" She looked like she would have liked to cut off my head and bury me, right there, next to the fountain. Clearly, this was not a person you want to mess around with, I thought, as I took several steps back.

Just then, Latsis turned to Clotro and apparently barked out some orders in an extremely foreign-sounding language. I mean, it sounded very weird. Immediately, Clotro whipped out her bulb-shaped thing again. It spun in her hand again, as it did the day before, while her fingers worked the buttons, or whatever was going on there.

Merle spoke to me. "Ken, we are losing time here. And time is very valuable to all of us, for our time here is limited. You must be made aware that this is not a joke, and we are very serious." He raised his arm and pointed up to the sky with his index finger. I followed his point, and once again I saw the black triangular kite-thing. "Do you see it, Ken?"

"The black kite? Yes." That struck me as odd. It was there again, in the same spot as yesterday.

Merle turned and nodded to Clotro. She waved her bulb in a sweeping motion in front of her, and a great rushing wind came up from the direction of the meadow.

Leaves and grass and twigs blew across the field towards us, smacking and stinging me in the face, while the trees whipped around and bowed down, all around us. Struggling to keep my feet in the violent wind, I looked out across the open space.

# CHAPTER
# 10

That was, without question, and by far, the single most astounding moment of my life up until that point. Everything seemed to stand still, crystallized in the moment. It felt like a giant window had opened, and I had the privilege of a stunning glimpse into a hidden world-- a hidden universe-- that had been there all along. I fell to my knees, struck silent and mouth open wide. My entire body quivered involuntarily at the astounding vision before me, maybe 100 yards away, out in the center of the meadow, where a moment ago nothing existed other than grass and dandelions, grasshoppers and butterflies, and open space.

It was, without any doubt, a gigantic black triangular spacecraft, probably 150 yards wide- half as wide as the meadow itself- and maybe 35 yards in height, floating only about 10 yards off the ground. I knew the ship was really there, because I could see the huge triangular shadow on the grass below, and it blocked most of the trees across the meadow out of view. The great wind passed as suddenly as it came, and the humongous craft floated in complete silence, slowly rotating with a gentle wobble, exposing in turn each of its three smoky matte-black metallic side panels, each one a perfect rectangle, and perfectly flat. I caught glimpses of the top and bottom of the ship as it rotated and swayed. It appeared to be perfectly flat on top, with the exception of some small protuberances here and there. The bottom was basically flat as well, with what appeared to be four rounded orange lights, or glowing ports. One large port was in the center, and a smaller one was at each of the three corners. It looked like there were additional, unlit ports, as well, but I wasn't sure. The ship was shaped like an equilateral triangle, or at least close to it, as far as I could tell. I wouldn't guess that something like that could float, let alone fly. But there it was, right in front of me, as plain as the leaves on the trees! Suddenly, and without warning, the muted side panels erupted in a wild multitude of brightly flashing, scintillating lights... I had to squint and hold my hands up over my eyes just to peek at the tremendous display, which seemed to out-shine the brightness of the morning sun. As bright as it was, I still couldn't help but to try to look at it, as best I could. The lights reflected off the meadow grass and the surrounding trees, cycling through every color imaginable *and* unimaginable. I stood there, stunned at the awesome kaleidoscope of power and beauty, as it blazed before me. I think I was too thunderstruck to panic, or even move, let alone run away in fear. I didn't want to miss one single moment of what I was witnessing.

That entire, amazing vision was seared into my consciousness, forever. In my mind's eye, I can still see every detail, as if the ship is still floating right before me. Even now, as I write this, I have goose bumps, as I always do when I relive that very moment.

How fortunate I am, born here on Earth at this time in our planet's development, to have witnessed firsthand such an amazing technological marvel; an inter-galactic machine untouchable by Earthly technology for who knows how many hundreds, or thousands, of years off into the future. At that moment, I couldn't even *dream* of ever being *inside of* anything like that.

Nearly overwhelmed by the sight, I began to slump down a bit, and I craned my neck forward to stare at the amazing craft, in all its dazzling glory. As I watched, a nearly inaudibly deep hum began emanating from the ship. It was a much different sound than the droning buzz I had heard just a short while back—this sound was deeper and less horn-like. Suddenly, in the blink of an eye, the lights shut down completely. The side panels reverted back to the solid, smoky black they were originally. Next the humming stopped, and the triangle stopped rotating and wobbling. It froze for several long moments, completely silent and motionless, so that all I could see was a giant, solid black rectangle, suspended low above the meadow. Then the entire ship began to fade away, right before my eyes. In about five seconds, it had literally just dissolved out of view, as if it had never even been there at all.

Then another great blast of wind tore through the area, almost tumbling me head over heels into the meadow with a powerful suction, and I looked back up towards the sky from my hands and knees.

The kite-- or what I thought had been a kite-- was gone. Suddenly a very deep feeling came upon me, like a huge feeling of loss had just opened up in my heart, seeing that empty sky up there. O.K., so it was definitely not a kite! I was definitely quite convinced at that point! I climbed back up to my feet, and for several moments I stood there, stunned, stretching my neck upwards, peering all over the sky, looking back to the meadow, and back to the sky, waiting-- hoping, actually-- for a return appearance.

Then I suddenly realized that my four acquaintances had vanished along with the ship. I stuck around another twenty minutes or so, looking and hoping for my new friends or their ship to return, to no avail.

# CHAPTER
# 11

As it dawned on me that neither the ship nor my acquaintances were returning, it also occurred to me that I should have taken some pictures, with my cellphone. Here the ship was *right in front of me*, plain as day, and I never even thought about taking a picture! Then I realized that I didn't even *bring* my phone! I was so mad at myself!

Heading out on a bike ride, without even taking my phone! And then not even thinking about it anyway, when the opportunity for a photo was right there! I pounded the heel of my right hand on my forehead, frustrated that I had completely and utterly blown the most amazing opportunity of my life.

Hopping back on my bike, I pedaled home probably faster than I had ever pedaled before. After I got back on the street, I was flying across intersections, blasting over high curbs, weaving between cars and pedestrians, careening along like a bicycle messenger on a crazy dare. Even though I was so teed off about not getting a photo of the ship, or of my mysterious friends, I knew that I had to call my family, my friends— everybody-- to let them know what happened.      I literally crashed my bicycle into the back yard, and ran into the house like a madman.

I threw open the door, ran to the phone and called my parents to tell them what had happened. I was so relieved when the phone was answered on the first ring! It was my dad.

Let me just say, I didn't get very far with my space alien story. I tried to tell him about the trips to the forest and the four people that I was convinced were aliens. Before I could even get to my face-to-face encounter with their spaceship, I could see where the conversation was headed.

"There are no space aliens out in the forest preserve, Ken. There are a lot of strange people that hang out there, though, that you might *think* are space aliens." That comment greatly amused him.

That's how the rest of the conversation went. No matter how adamant I was that I saw a huge spaceship, close enough to throw a rock at, he kept getting back to his point.

"I don't think you should go back there, Ken. I'd just stay away from there."

And that was about it. That call basically went nowhere. And to this day, my dad still doesn't want to discuss that part of my Enlightening. After that reaction, I realized that I might want to rethink my plan of calling everybody I knew. I didn't want to alienate all my friends and family, in one big rush of me being excited.

There was also a bit of a problem, truthfully, in that my new friends, who I assumed must definitely be space aliens, really looked like normal humans. Merle had a bit of an unusual look about him, I suppose, which I could never really put my finger on, exactly—I think he was unable to completely mask his facial bone structure, or head shape, or something, as he assumed his Earth-human disguise. But otherwise, there was no way to tell, whatsoever, that Merle was any different a person than you or I.

I later learned that Merle's planet was selected as the source of the mission partially because there was a considerable physical resemblance between Merle's people and Earth's people, as well as a similar gravitational field on each of the two planets. Similar height, weight, body structure and facial structure, for the most part. Some of the other alien species were too short, or with too large of heads, or too tall, or too used to other gravitational conditions. Especially since Merle was going to be spending a lot of time on-planet, and interacting strongly with Earth-people, myself included, it was better that any disguise didn't need to be dramatic and unwieldy.

As far as the three girls go, they looked just about perfect, really, even considering some of Atropha's rough edges. Each of them would definitely turn a few heads down here on Earth. They looked like perfectly normal, albeit rather exotic and beautiful, Earth girls. I was surprised when Merle once told me that the girls' Earth-people disguises were quite different than how they appeared in their own real life. He implied that the true appearance of the girls was a little outside of the box. That was intriguing, but also, I have to admit, a little unsettling.

After a while, I got to thinking that these girls were basically using avatars, almost like what you might use in an online game or something. So that added another weird dimension to the whole thing for me. I almost wish Merle hadn't mentioned anything about what they really looked like. Still, it's always good to get a full picture of the situation, if you could say that knowing that these girls were basically picture-perfect avatars was getting a full picture of the situation. Well, you know what I'm saying, I think.

After I hung up the phone with my dad, that day, I realized that it might be hugely beneficial to take some photos, just to document these individuals, even though it might be a tough sell considering their perfectly normal appearance. I hoped that, if all went well, I might get the money shot-- the spaceship. So after I hung up, I ran out the back door, quickly adjusted the chain on my now slightly banged-up bike, wiped the grease off my hands, and headed out to a big electronics store to see what they had in the way of small cameras. I bought a couple of mini video cameras and a tiny still-picture camera, as well, with my credit card, even though I didn't know how I would make the payments, on my small paycheck from The Enterprise, which is what we called the R.H. Enterprise, for short.

When I got home, I spent the rest of the day rigging up the cameras inside my clothing and my backpack. Along with my phone, I was going to secretly capture some photos and video footage that would prove to everybody that I hadn't lost my mind.

After I had everything set up, my mom called. Obviously she had talked with my dad. She tiptoed around the "space alien" angle by referring to the "strange people I met" at the forest preserves. I had to reassure her that I was OK. I told her that I agreed it was probably best to stay away from there.

I went to bed that night secure in my thinking that I would take some great photos the next day, at the forest preserve. I figured that close-up photos of the triangle would be a world-wide sensation, once they got out, if I could get another chance.

# CHAPTER 12

That next morning, I got all my hidden cameras set, and, pulling my bike out of the garage, headed out extra-early. I stopped in my tracks as I realized that I was hearing that deep, fuzzy droning sound again. This time it stopped almost as soon as it began, and I rolled down the driveway towards the street in great anticipation. I immediately noticed that music was coming from somewhere. It sounded like Pink Floyd, but I wasn't sure. I looked out and saw a car parked out at the curb in front of the house. It was a regular looking little hatchback-type car, and Merle was in the driver's seat. He was listening-- very intently-- to a song which he later told me was "Brain Damage", by Pink Floyd, from *Dark Side of the Moon*. Tapping his fingers on the armrest of his chair to the beat of the music, a smile of sly satisfaction was spreading across Merle's face. He turned his head in my direction, and our eyes met.

I hit the brakes-- hard-- and hopped off the bike. It fell from my grasp and landed on the driveway in a clattering heap.

Merle turned off the music and came bounding out of the car like a cat, to help me pick up the bicycle. It seemed like he was there in an instant. "Let me help!" he said as he grabbed the bike and stood it upright.

"You have a car?" I asked.

"Sort of," he grinned. "Sort of a special model." Well, that was an understatement, as I came to find out. "Come on, we can put your bike in the back." And he rolled my bike over towards the back of the car.

I sort of panicked at that. This was the ultimate "ride with a stranger". I was scared to go in there, and I sure didn't want my bike in there, either. Quickly I looked for a way out. "No way, Merle. That car is too small."

He laughed out loud at that. "This car? I don't think so! I could put *twelve* bicycles in here. Come on, have a look!" And with that, he lifted the back hatch.

I sure didn't expect what I saw next. With the hatch shut, and viewing the inside of the car through the closed windows, it looked just like the inside of a regular car. But once he opened the hatch and I poked my head in there, an entirely different scene unfolded. It was *huge* inside! He wasn't kidding about the 12 bicycles. I could probably have gotten 30 bikes in there.

And it wasn't the inside of a normal car I was looking at. It was a full circular shape on the inside. The ceiling, walls and floors were covered in a sort of glowing metal, which appeared to be self-illuminated with a soft, bluish-white light. Panoramic windows rimmed the entire craft for near total visibility of the outside surroundings. There were a few blinking or flashing gauges or whatever in different locations, but surprisingly few, actually. Several were on the sizeable central console, which was under the front window and extended out to within easy arms' reach of the driver and front-seat passenger. Those two seats were the only visible seats in the entire craft. In the center of the vehicle, several yards behind the seats, was an orange-colored cylinder, metallic-looking and yet mostly transparent, maybe five feet in diameter, which stretched from the floor to the ceiling. There was a rearview screen that was in the top of the front window, but you could also turn around and look right through the orange tube as if it almost wasn't even there. All in all, the entire scene was basically like what you'd expect the inside of a spaceship to look like, I think. But, of course, that was because the car *was* a spaceship, more so than it was a car, as I came to find out.

"Sort of like the "Tardis", huh?" Merle was grinning widely at me. At the time I didn't understand the "Tardis" reference. I'd never really watched any "Doctor Who" before that. "Can we load the bicycle in?" Merle asked. "I'd like to get going."

Well, what choice did I have, at that point? I picked up the bike and was about to put it inside, when Merle put up his hand to stop me. "Hold on," he said. "There's one more thing."

"What?"

"I'm going to need you to leave all your cameras at home. No photos or video. Those are my orders," he said. "That was my agreement."

"I don't have any cameras," I lied. Boy, do I regret saying that.

Merle looked at me, with the look my mother might have given me when I was seven, after catching me in the cookie jar before dinner. "Ken, you can't get in-- or out-- without passing by the landlady in the kitchen, do you know what I mean? Please return the cameras to the house. I'll wait. If you want to put your bike back in the garage, that's fine. You won't need it."

He didn't seem angry, or even judgmental. In retrospect, that hurt. Almost like he just assumed I would lie about it, like I was a young child. I guess that's a pretty accurate assumption and description, though.

I asked Merle about the landlady in the kitchen reference once, and he mentioned something about some Russian novel. Well, that went right over *my* head. The amount of knowledge Merle had about Earth culture was unbelievable, especially considering that his home planet was 97 light years from Earth.

Sheepishly hanging my head after being caught in the fib, I retreated back to the garage with my bike.

"And leave the phone also!" Merle called after me. Damn, he didn't miss a trick.

I already knew better than to test Merle. And after I had gotten a good look at the inside of the car, I never considered not getting into it with him. I don't remember being afraid, or even concerned about it, in the least. I just felt like I had to get back into that car, or ship, if I had any opportunity at all to do so. I was very focused on that idea.

So I put the bike in the garage. Then I hurried back into the house and dumped all the camera gear, as well as my phone, on the kitchen table in a heap, so I could hustle back to the car. I never did get even a single photo, the entire time Merle was here.

When I first headed out that morning, I didn't have anything turned on yet. I didn't want to waste the batteries, so I was going to fire everything up as I approached the forest. I never got the chance. Nor did I ever try again. And I'm fine with that. I know what I saw, and what I experienced.

# CHAPTER
# 13

When I got back to the car, Merle was back in the driver's seat, and he was again listening to Pink Floyd. It was now the song "Eclipse". He motioned for me to come inside. I opened the door-- it was exactly like a regular car door on the outside-- and sat down. I looked around the inside of the vehicle in amazement. The seat looked like it would be hard, like hard plastic or metal, but actually it felt incredibly comfortable once I sat down. It contoured to my body perfectly, and gave a little-- but not too much-- with every move. It was the most luxurious and comfortable chair I had ever sat in, by far. It almost seemed like it was alive, it was so responsive. Sometimes it almost seemed like the entire ship was alive, really. It's hard to explain.

Merle smiled at me, and, concentrating on the music, tapped his finger on the armrest, while silently mouthing the lyrics.

The big orchestral climax faded out into the sound of a beating heart, which I had never really noticed before in that song. As the heart continued its beating, Merle turned to me and spoke the spoken final lyrics, perfectly mimicking the British cockney accent on the original record, "There is no dark side of the moon, really. As a matter of fact, it's all dark." Then the song ended, and Merle chuckled. I figured that he must have found the ending part to the song to be funny.

He turned to me again. We were still parked at the curb. "I just love that album," he told me, "*Dark Side of the Moon*. It was released in 1973." He turned to me and smiled.

"You know, that album played a very large role, believe it or not, in my getting this mission.

"All the space-faring peoples in this part of the galaxy were very fortunate to have such a close view of the development of the Earth, especially at such an important era in Earth's history. We on Akeethera had been observing your planet for a long time.

When you started detonating atomic bombs, it really got everybody's attention. Then you had two massive World Wars, not to mention many other wars and skirmishes continuously going on somewhere, all the time.

"And you were just taking your first tiny baby steps, sending up satellites, orbiting the Earth, and then, finally, landing men on the moon. I remember how interesting it was, for me especially, observing your planet's development as it put its first feelers out into the surrounding universe. Watching Neil Armstrong take the first steps onto the surface of your moon was one of the real formative moments of my youth, believe it or not! I remember watching it with my mom, while we were having breakfast, from the same news feed that was being broadcast back to Earth.

"I remember my mom saying that we were probably seeing the feed even before the people on Earth were seeing it, because she had pulled some very big strings to get us linked in to a very fast transmission network, just for the event. My mom had decorated the room with Beatles posters and other Earth-style paraphernalia, to get us in the proper mood, she said. She even had a lava lamp going.

"My mother and I were amazed that the people of Earth were operating with a very limited and incomplete view of the universe, due to some basic misunderstandings underlying your Relativity theory. That fascinated us as much as anything, I think.

"A lot of people on Akeethera, at that time, followed the various Earthly developments from our news feeds, and so forth. Many still do! Being young, as I and my friends were at the time, we especially enjoyed the popular American and British music of the day. It was very different than Akeetheran music, and the uniqueness appealed to us greatly. There was also a rawer edge to your music, which also appealed to us, as we were able to more clearly imagine some of the challenges the people of Earth faced, that we as a society had not faced for more than a millennium."

Merle told me that he had really hooked into Earth music through his uncle, who had run a powerful supply and transport ship for several years, on the final leg of the Akeethera-to-Earth route. Merle's uncle had a front-row seat, basically, to the whole British Invasion music scene, as it was happening. He never was on-planet himself, but he still had to be fully inoculated against Earth pathogens, since he got plenty of exposure to the various travelers he met. These travelers were shuttling back and forth between Akeethera and our solar system, and on occasion the Earth itself, "for scientific purposes, or for pleasure, or any combination of the two", as Merle had said with a grin and a wink. And they were happy to share their findings with Merle's uncle. So when Merle's uncle came back to Akeethera, he passed along his love of Earth music to a young Merle.

"I, myself, absolutely loved the Beatles, even though they broke up just after I started listening to their music," Merle told me. "I remember I had a huge Abbey Road poster on the wall of my room, in my parent's house. I absolutely loved that Abbey Road album."

I sat back in my seat, totally astounded by Merle's comments, while he continued speaking. "Many people on Akeethera were disappointed when the Beatles broke up. I remember seeing it in the news, at the time it happened, and I remember how sad my mom was, because she liked them, too. Even still in 1973, I remember wondering if they might get back together, because just as I was starting to get into the whole American/ British rock music thing, and particularly the Beatles, they broke up.

"Meanwhile, I was listening to a lot of new music from Earth, and hoping to come across the next big thing. Of course there were a lot of other great bands out there, at the time. I was somewhat aware of Pink Floyd, also, and I thought they were very interesting. And then I heard *Dark Side of the Moon* for the first time." He paused and shook his head, as in amazement at its greatness. "There is so much pain, confusion, fear and doubt on that album, which greatly appealed to me, at the time, as an artistic statement from Earth. I felt it was relevant even in my own society, as far removed from Earth as that may be." Merle mentioned that the album, on one level, referenced Syd Barrett, who I had never even heard of before. He was actually the artistic leader of the band in their early days, but Syd was no longer in the band, well before they recorded *Dark Side of the Moon.*

"The album, to me, really touches on the *universality* of the feelings of fear and anxiety and isolation," Merle said. "I love how the band used the imagery of the eclipse of the sun, and the dark side of the moon, as they relate to becoming separated from the ties of connectivity that bind us all together, and to illustrate what a profound magnitude of loss that would be. Also, how easy it is for any one of us to fall into that isolation and separation, even in a bustling, connected world. And also, who really *are* the crazy ones?" Merle smiled, clearly enjoying his memory of the artistry of the album. "At least, that's what it means to *me*."

Merle had already made enough cultural references that I realized that he knew a thing or two about the ways of Earth. And after all I that I had already witnessed, I was starting to realize that it was just like Merle to surprise me with the unexpected. So I accepted his comments on *Dark Side of the Moon* in silence, without even nodding my head. Merle continued speaking, slowly, as if he was trying to formulate his thoughts as he went along.

"I remember, as a beginning student of Earth, how interested I was, learning about your planet and its history, from the research of people from my planet, and others, who had been studying your world for thousands of years; if not a lot longer." He winked at me as he said that. "I tremendously enjoyed learning about so many of the great thinkers and artists in your planet's history, obviously many more than one man could study, even in the longest lifetime." He slowly shook his head, deep in thought.

"And then, as a young student, I became so very interested in this collection of songs, by this somewhat obscure, modern British rock band. Just like a teenager!

"Even though I could speak English relatively well, it was still difficult for me to understand all the lyrics of the songs I was listening to. I remember that, as a young member of the academy, I officially requested a copy of the lyrics be sent to me. I don't know what made me think anybody would actually get them. At the time, we didn't even have a full-time man on-planet, and the Committee on Western Earth Culture had to go through another source, from one of our partners in another system who had a person here. There wasn't any internet back then, of course, with all the lyrics on-line, so she actually purchased a copy of the album in order to get the lyrics in writing, so she could have a copy of the lyrics sent to me, from right off the actual album sleeve. Ha! A lot of trouble to go through! I'm afraid the Earth culture representatives on the committee thought I had... taken leave of my faculties, I think is how you say it. But after a few days, I received a copy of the lyrics, transmitted 97 light years, across several sectors of the galaxy, just so I could understand it better!" He laughed out loud at that memory.

"One day several years ago (depending on how you look at it), while I was still on Akeethera, I was talking to Risdef Droynom, in a hallway of the academy. He was the head of Western Earth Culture Studies for the committee at the time. Risdef is a wonderful man, and he was the one who pulled the strings to have the lyrics sent to me. More than anyone, he thought I was... barking up the wrong tree... and maybe wasting valuable resources, with my interest in *Dark Side of the Moon*. He might have thought it was a pointless teen-age obsession, I'm afraid! Yet he still agreed to help get me the lyrics.

"That day in the hallway, we were talking about many things. I remember we were discussing the influence of some of the greatest Russian authors on Western literature. Somehow, that led to us talking about Charles Dickens, Emerson and Thoreau, I believe; Mary Shelley and Rachel Carson, also. Then we were talking about Bob Dylan.

"Finally, he put his arm around my shoulder and asked me if I was still listening to *Dark Side of the Moon*. Of course I said yes. I'll never forget how he reached into the large bag he was carrying, and pulled out a package. As soon as I saw it, I knew what it was. It was an actual vinyl album of *Dark Side of the Moon*. 'I thought you might enjoy this,' he told me. Risdef had that album shipped all that way-- on a series of four different very fast ships-- just so I could have it. Even though I didn't even have a record player to play it on! I was so stunned I didn't even know what to say!

"I was looking at the album in my hands, and stammering as I tried to think of something to say, when Risdef told me that I had gotten the commission for being one of the on-planet contacts for this mission! That, right there, was the happiest moment of my life up until that point, without question... I had worked so very hard to get to this commission! Years of deep study and immersion, to get to that point!

"That night, when I got home, I was so excited. I put the album on the table, opened it up, and poured myself a little fluzle, to celebrate. I played the music over and over, from another recording of it that I had previously gotten through the communication network, while I just sat and stared at the *Dark Side of the Moon* album, direct shipped all the way from Earth!"

"Fluzle?"

"It's sort of like the Akeethera version of champagne," Merle said. He then let rip with a big, full-throated laugh, which again took me by surprise. He held his slim belly and laughed some more, as he remembered the moment. "Even though I might have had a slight advantage, I was one of only three people chosen, out of more than 2400 candidates, for one of the on-planet spots on this mission, which I still can barely believe! And that vinyl album, to this day, is my most valued possession."

At the time, I didn't know what Merle meant by his "slight advantage", but eventually I did come to find out what his unique advantage might have been.

# CHAPTER
# 14

I came to find that although Merle's knowledge of science was light-years beyond that of present-day Earth, he was especially fond of Earth culture, especially music and art of just about any variety. Merle once explained to me that his planet's history, like Earth's history, included a man who discovered gravity. His planet's history, like Earth's history, included a man who discovered that their planet revolves around their sun.

Likewise, any other scientific achievement that we celebrate on Earth had already been discovered on Merle's planet, many millennia ago. There was always a certain sense of "been there, done that" when it comes to Earthly scientific advances, from Merle's vantage point, I guess is what he was trying to say, but he was very tactful in saying it.

However, art, unlike science, is unique. Nobody on Merle's planet had ever performed "Manic Depression". Nobody on Merle's planet had ever written "100 Years of Solitude". So Merle was understandably drawn towards the unique cultural perspective that art can provide. And I know that Merle's strong cultural investment was very important to the success of the mission, even though it was really more of a science mission, at least on the surface.

As it turns out, most of Merle's research into our western culture was sort of rooted in the 1970s and previous eras. That was because of the great distance between Earth and Merle's home planet, Akeethera, which I am spelling phonetically, based on what Merle called it, with the accent on the "double e", and very little emphasis on the final "e". When he started preparing for his quest, back on Akeethera, it was the

1970s on Earth. As I had already mentioned, the transmission relays, built by the various inter- galactic societies and scattered throughout the galaxy and beyond, are able to accelerate the speed of information transfer across great distances, so Merle was able to stay relatively up to date, in spite of the great distances involved.

As Merle eventually traveled across the galaxy in his transport ship, accelerating to velocities well beyond "light speed", he aged quite slowly in comparison to the time frame of Earth, so the decades flew by on Earth, as he travelled. It wasn't until Merle was already on his way to Earth that I came into the picture as a potential contactee, after I rescued Magu from that runaway car. During his trip, Merle spent much of his time researching various aspects of his mission, and he no longer spent much time listening to the latest Earth music. So I always kidded him about being stuck in the past with his Earth-musical tastes. He said he liked some of the modern music, but he just didn't have much time to study it, like he did back in the day. I think maybe he was just trying to be diplomatic by saying that, though. You don't "study" popular music. You enjoy it!

In some ways Merle was just like my own father, or grandfather, even, listening to the oldies station… But he loved a lot of different music. Sometimes it was classical music-- Mozart or Beethoven or Tchaikovsky or whoever, or maybe blues, country or jazz-- he loved Django Reinhart and Thelonious Monk especially. I remember in particular one night when we were in his car/ ship, and he was dropping me off at the house after the most eventful day of my life up until that time. We parked at the curb and sat in silence, listening to "'Round Midnight", much like we sat listening to "Eclipse" that first day. The fireflies were out flashing around the yards and the little park across the street, and Merle thought it was the perfect ending to the night, watching the fireflies, which completely fascinated him, and listening to Thelonious Monk. The song--'Round Midnight"—was only a few minutes long, but Merle was completely knocked out by it. I remember him saying that it was as beautiful piece of music as he had ever heard. He loved quite a variety of different music and bands, though, often commenting on the lyrics as well as the instrumentation.

Anyhow, after he finished with his critique of *Dark Side of The Moon*, Merle turned to me again and said, "There is one more thing we need to discuss, before we go anywhere, Ken. We are lucky that your father and mother were so skeptical, about the girls and me, and the spaceship, when you called them last night."

I was shocked that Merle knew about my phone calls with my mom and dad, the night before. All I could ask in response was "Lucky?"

"Yes, very. The single most important request I have of you is that you do not reveal any more details about my visit, to anyone. Keep this… under your hat. Is that right to say, even if you are not wearing a hat?"

"Under my hat? Uh, sure. That's right. No, I don't have to be wearing a hat for you to say that."

"That's what I thought, but it didn't sound right. I wanted to make sure."

"You were right."

"Good, good. Anyhow, you are not to speak a word of this to anyone, Ken. Nobody else must know about my visit, at least until I'm gone. After that, all bets are off, as they say. After that, you can tell anybody anything you want. But until then, that is the prime directive."

"All right." I agreed with the request immediately, since I had just seen two examples of Merle knowing things about me that I hadn't expected. "I won't tell anybody. I know you can't get past the landlady, anyhow, right?"

Merle laughed at my usage of his landlady reference, since I barely even knew what it was all about. "That's right, Ken. That landlady seems to know it all, doesn't she?" He smiled at our little ongoing charade. He turned back to face the road, and he touched a small box-like protrusion on the screen, that I hadn't noticed before. After he touched the box-like area, a rippling bluish haze seemed to envelop it.

As the car began to move, Merle smiled again. "OK, let's see what this baby can do!" I was already discovering that Merle loved pulling out trite American sayings. He knew a million of them, it seemed, and always got a kick out of using one correctly in context. He peered around to look for other cars, and I was alarmed to notice that there wasn't a steering wheel or any obvious means of steering his "car", nor any pedals on the floor. Nevertheless, we smoothly pulled out into traffic, which was very light at that hour, and we were on our way. At first, I was pretty nervous that Merle appeared to be just sitting there, watching everything, as the car seemed to drive all on its own. But Merle assured me that he was, in fact, directing the vehicle.

From the outside, Merle's "car" looked to be a perfectly normal smallish American car; maybe a bit of a "beater", complete with patches of rust and scratches in the paint. Nothing fancy whatsoever. Even with the windows rolled down, anybody looking from the outside would see Merle driving, and actively steering with his hands on the wheel, just like any other driver. But once you got inside and sat down, you saw an entirely different vehicle, with Merle directing the movements of the car with his mind, apparently. Merle once told me that at any instant, if circumstances necessitated, we and the entire car would be safely transported to the loading bay of the main triangular ship. I secretly hoped that would happen, assuming that I would be transported there also, which obviously would be the thrill of a lifetime-- even more of a thrill than simply seeing the ship up close, as I already had.

As Merle pulled out into the street that morning, handling the car like an old pro, somehow, he told me that he'd like to pay a visit to the local park, "Streamside Park". It was a Saturday, and there would be a lot of activity out there as the day progressed. I could tell he was really looking forward to it, and I agreed to go there. I used to go to that park, with my mom and dad, when I was a little kid. I remember that I used to like to chase grasshoppers and butterflies, and I'd go down by the stream to see if I could see any crayfish in the water. Later, when I was older, I used to go there to play baseball, football, and basketball. In the winter there was even an ice skating rink there, where we used to play hockey sometimes.

At that moment, however, I wasn't thinking too much about those days of youth. I was just stunned at how I ended up to be sitting in that car-- or whatever kind of contraption it was. Then, I started thinking about all the time I wasted the night before getting ready to take videos and photos. And then I realized I had a great opportunity to ask some questions of Merle.

"So is this thing is really a spaceship, or what?"

"I guess you could say that."

"Is this the ship you took from your home planet?"

"This? No." Merle chuckled at that one. "This is just my local transport ship, basically, and one I can drive on the streets."

"So you came here in that giant triangle?"

"Yes. We've been here for over five years, now."

"Did you really travel 97 light years to get here?"

"Well, my home planet, Akeethera, is about 97 light years distant. But we actually travelled more than twice that distance on my way here. When you go above the space/time ratio, you usually have to be careful about your route, for safety. Plus we made stops along the way at 12 other systems, to assemble the crew. This is a multiple-system mission, and the black triangle is both the bus and the base. Roughly 43 percent of the ship's population is from Akeethera, with the balance spit between the twelve other systems, along with a variety of visitors from many other systems, as well."

"Wait! Did you say you went faster than the ratio of space to time?" At the time, I thought Merle was just pulling my leg, even though he had primed the pump by discussing the Lorentz transformations multiple times, already.

"Well, you would say 'faster than the speed of light.' But hopefully you already know that the 'speed of light' is actually the active ratio of space to time in the universe. So we took a roundabout path, and made a few stops along the way, which added some distance. Also, on a mission like this, there's the time management side of things. We wanted to take some time along the way for study and learning purposes, before getting into orbit, but we didn't want to take *too* much time. So that's how we did it."

"Where are your three girlfriends?"

"That's hard to say. But they're never really very far away, I can assure you."

"Did you travel here with them?"

Merle chuckled again. "No, I didn't. They arrived here long, long before I was even born. Even though, in a sense, I've probably spent half as much time here than they have, by now. Those three women are time-savers."

"Time-savers?"

"That's right. They have their own very fast ship, and they spend most of their time off the planet, or more accurately circling the planet, traveling at relativistic speeds—or beyond relativistic speeds, even-- relative to your planet's time frame. Anyhow, by maintaining that type of velocity as their standard operating velocity, they are able to age extremely slowly. They save time, in other words, relative to the time frame of the Earth. So one second of their time may be an entire day on Earth. Half a minute of their time, versus one month on Earth. Six minutes of their time, and an entire year may pass by, on the Earth. One hour, ten years. That sort of thing."

"Wow." I quickly worked out the numbers in my head. "One thousand years, four days?"

Merle laughed. "Yes, as incredible as it may sound, that is certainly possible. Of course, when they come down onto the planet, their time passes at the same rate as anyone else on the planet. So they are not saving time while they are actually down here. That's why they can be a little impatient at times- especially Atropha." He turned to me and winked, as he mentioned Atropha.

"She seems pretty fierce."

"Uh… yes. Ha ha! Yes, most definitely, Atropha can be pretty fierce! Not one to be trifled with, I'm afraid… Many have learned that, to their dismay."

"And what's the deal with the thing that Clotro has?"

"Oh, that! Fast and bulbous, isn't it?" Merle laughed. From time to time Merle would amuse himself with some pretty unusual comments.

"I guess so... But what is it?"

"Well, I call it her device. It's like a transmitter, or a remote control, or a computer. I wouldn't mind having one of those little gadgets, myself."

"So how long have they been here, exactly? A long time?"

"Well, time, as they say, is relative. That is very true. But they first arrived here about 27,000 years ago, in Earth years. So, in their timescale, they haven't been here very long at all. They've probably spent about ten years of their own time here."

"What?!" I couldn't help shouting it out. "Merle, are you saying those three women have been here since the *Ice Age*?

"Oh, yes. Absolutely. Those girls have been around the block a few times."

"Crap, Merle. No way. *No way.*" I was really astounded by the thought. "Really?"

"Oh, yes. Those three are very interesting people." Merle typically referred to all races of beings from various planetary systems as "people", which sometimes confused me until I got used to it. "They've had some great experiences here, and they have some amazing stories to tell. They have been places, and seen things, that are really quite astounding. And they have personally encountered many extraordinary-- and ordinary-- inhabitants of Earth. I envy them, very much. I wish I could spend a lot more time with them than I have been able to. But of course, they are--"

"Time savers?"

"Yes. So between helping me, and helping my two other mission-mates who are elsewhere on the planet, they don't have too much more time left over, at the moment. In fact, I'm grateful they agreed to spend so much of their valuable time on this mission.

And they have freely given me access to much of their archive of information. I think that tells you something, right there."

"Tells me what?"

"It tells you that this project is important," said Merle. "Also, it tells you that they value this planet tremendously, as do I."

Merle was focusing on the road the whole time we talked, stopping and turning smoothly and naturally, as we discussed the three time-savers.

My mind reeled at all the things that happened during the time the time-savers circled Earth, if Merle was telling the truth; things they may have been witness to. They saw the end of the Pleistocene era; the end of the last ice age. Mammoths, mastodons, saber-tooth tigers, cave bears. Neanderthals! The beginnings of civilizations, farming, and all the major religions. So many things! I was certainly viewing these women in an entirely different light! I got right into the biggest question I could think of.

"Merle, did the time-savers ever meet Jesus?"

"I'm afraid I can't really give you an answer to that, Ken."

"Why not?"

"I'm not allowed to discuss specific religious beliefs of your planet."

"Why not?"

"I have rules. No photos or video. No discussion of specific aspects of religious beliefs-- yours, mine, or anyone else's. And there are many more rules I have to abide by, also."

"How would anybody know what we talked about?"

Merle laughed out loud at that one. "Once again, my friend, I must remind you; you have to pass by the landlady in the kitchen. The same is as true for me as it is for you! I have to abide by the rules set out for me, or I'll have failed my mission, and I'll be pulled out, just like *that*." And he snapped his fingers.

"*Who* will pull you out?"

"Well, there's a consortium of local planetary systems that are sponsoring our little adventure; providing support, *as well as oversight*."

He turned to look at me and he sort of raised his eyebrows and tilted his head upwards, looking up to the sky, as he said that part about the oversight. "This thing has taken many years of discussion and planning, with many different groups of people that have an interest in the situation. Just about every mid-level type society within this entire mega-quadrant of the galaxy is involved. Plus, there is some higher level input, even from inter-dimensional portions of the galaxy, and from outside of the galaxy as well, I would imagine."

"Higher level input?"

"As you've seen, Ken, I'm employing certain technologies that are very far advanced, in comparison to your own. In the same way, there are societies that are as technologically advanced beyond me, as I am to you. And so on. Some of these people operate in ways that are very far beyond my own abilities, or even my own awareness. I'm quite sure that there is an element of that type of 'higher level input', as I referred to it. That input includes some higher level *oversight*, as well." He glanced over to look at me as he said that, and he raised his eyebrows a bit, again.

# CHAPTER
# 15

By now, Merle was preparing to pull into our curbside parking space at Streamside Park. A big black SUV raced up behind us, and the driver honked at Merle, impatiently, before racing past us. Merle turned to me and smiled. "I suppose some people think that we're not allowed to slow down to park, are we? And I had my turn signal on, too!" Although it seemed like our car, or ship, was way too large to operate on the street, let alone pull into a parking space, somehow Merle was able to parallel park as easily as you can imagine. Somehow, we squeezed into a normal-sized parking space between two other cars.

"How in the heck does this ship fit into that spot?" I asked Merle.

"Good question!" Merle seemed happy I asked. "This ship is actually oscillating very rapidly, at all times. Since we are inside the ship, and the entire ship is oscillating, we don't feel it, since we're oscillating, also. But the net effect is that, even while 'parked', we are oscillating at relativistic velocities, and therefore we've reduced our size, relative to the space/time reference frame of the Earth. So we fit. The camouflage system of the ship is able to alter or correct the visual appearances, so that everything appears to be perfectly normal. Not only to a person on the street, but also to us, inside the car.

Or ship." He smiled again. "I never know what to call it."

"It reminds me of the DeLorean from *Back to the Future*", I said. "Just way less fancy on the outside, but way more fancy on the inside."

"*Back to the Future*? What is that? And what's a DeLorean?"

"That was an old movie, sort of a sci-fi comedy, I guess, where this guy customizes a DeLorean—which was a brand of car-- and turns it into a time machine."

Merle grinned at that one. "*Back to the Future*, huh? I must have missed that one!

Although I did love *2001: A Space Odyssey*! What was the name of that computer, from the University of Illinois?"

I had no idea what Merle was talking about.

"HAL 9000! That's right!" Merle answered his own question before returning to my *Back to the Future* reference. "A time travelling car, you say? Well, this one can travel in different *dimensions* of space-time, but I don't think that's quite the same thing…"

At the time, I let the comment about the different dimensions of space-time slide past, because I was focusing on something else that Merle had said a few minutes prior. "So where are the other two on-planet people? You said that you were one of three people chosen to be an on-planet contact."

"Well, right now, one of them is probably on the base" (by that Merle meant the big triangular ship), "studying or getting some sleep. But, generally speaking, they're each spending their mission time in different countries of your planet. I can't really mention which other countries, specifically. But I'm the Western mission specialist, based in the U.S. Then we have an Eastern mission specialist, based in another country on the other side of the world. He's probably asleep on the ship right now. Also we have a Southern mission specialist, and she is based in the southern hemisphere, is all I can really say. Basically, though, each mission specialist is working with a single person from Earth, with the goal being to begin opening some eyes on Earth to the realities of Hyper Relativity."

"And why is that so important?"

"That is a really great question, Ken. Basically, what is the point of any scientific endeavor?"

"Well… I guess to advance human knowledge. Or, to advance the knowledge of whatever civilization."

"Good. True. I like how you're already expanding your thinking, Ken. But why is the advancement of knowledge a good thing? A lot of people on Earth feel that science often seems like a pointless advancement of knowledge, don't they? So what is the end game, as far as the scientific advancement of knowledge?"

I admit I had to sit there and think about that one for a while before I could come up with an answer. Even then, I wasn't too sure it was the answer Merle was looking for. It turned out that it was, though. "To help people?"

"EXACTLY!" Merle vigorously nodded his approval. "Exactly true! To help people.

That's what *should* be the end game, anyhow. Of course, sadly, there are exceptions to the rule, but in general, scientific advancements have a way of not only helping the individual, or not only helping the community, but hopefully helping society as a whole to live better, happier, healthier lives. That is true, even if the way to a better life is not initially obvious. That's the whole point of advancing science, and advancing human knowledge in general. Knowledge trumps ignorance, Ken. One simple idea can have a tremendous, positive impact on so many people—maybe even an entire planet. One idea can reverberate across centuries, or millennia, of time. One idea can spread across the galaxy, Ken, and beyond."

Merle looked at me for a moment, and sighed deeply, before he continued. "The civilization of your planet has reached a surprisingly advanced stage, to have not figured out Hyper Relativity yet, Ken. Maybe that is part of the reason you continue to wage war so damn incessantly!" Merle thumped his fist on the console in front of him, for emphasis, and I was a little surprised to see him express anger like that. "Many of your planet's countries have a great number of powerful and expensive weapons, many of which they fire quite liberally, while at the same time, these same countries struggle to gather the resources to feed and house their own people in safety!

"Also, you are polluting this beautiful planet, to an extreme degree. As you begin to explore your own solar system and beyond, a lot of observers outside of your planetary sphere are concerned about your intentions. You consume vast energy resources from your own planet, and you're driving an enormous number of Earth animal species to extinction, to your own obvious detriment. You are finding more and more ways to misuse science; not to help people, but to advance certain peoples' narrow interests at the *expense* of others, and at the expense of the planet itself. By extension, that hurts us all, anywhere in the galaxy, or beyond.

"Hyper Relativity is an opening of a window, basically, Ken. And once you open that window, a lot of other things come into play along with it. Once you understand the hyper-dimensional space-time continuum, and how it interacts with energy in the universe, you'll discover the shadow universe which gives rise to all the 'dark matter' and 'dark energy' you'll ever need. That kind of knowledge leads to new ways to interact with the universe, for the betterment of people. It's the shattering of old boundaries!" With that, Merle thumped his fist on the console in front of him, again, for emphasis. He paused a moment to gather his thoughts before placing his hand on my shoulder, as if to make sure he had my full attention. He peered into my eyes, intently. "How would you like to understand the true large-scale physical structure of the universe?" he asked me. "How would you like to understand the double-slit experiment?"

"The double-slit experiment?" I imagine that my eyes must have gotten as large as saucers at that one. I had been greatly interested in the double-slit experiment since I first came across it in high school Physics. The experiment was famous since the early 1800s for demonstrating that light was a wave phenomenon, as well as acting as a particle. The experiment gained an even greater air of mystery when it was performed with electrons, back in 1961. Since then, larger and larger units, including atoms and shockingly large molecules, have followed suit.

Basically, it goes like this: Pass light, or an electron, or a molecule, under the right conditions, through a single slit in a screen, and it passes through the slit in a familiar diffraction pattern, impacting a panel behind the slit in a pattern that looks much like a somewhat larger, fuzzy slit. Add a second slit, though, and instead of passing through in two fuzzy slits, or two simple diffraction bands (as was originally expected), the light, or electrons, or molecules, form a more intense diffraction pattern, with the addition of a multitude of additional interference ridges. These interference ridges are the hallmark of waves interfering with each other, creating brighter areas where the "crests" of the wavelengths intersect, and vacant areas where the "troughs" of the wavelengths meet.

That is why electromagnetic energy is considered to have a "wave-particle duality." The electromagnetic energy, such as light, clearly impacts the panel behind as individual points- i.e. particles. However, the interference patterns seem to prove that the light is also behaving as a wave, paradoxically.

Things got even stranger when the same results were discovered with electrons, therefore indicating that electrons travel as waves, as well, in addition to obviously being discreet particles.

Things got far stranger still, however, when a device was constructed that shot electrons *one at a time* through the dual-slit screen. Shockingly and incomprehensibly, even when fired one at a time, with clear and full intervals in between shots, the interference pattern still appeared. How can an interference pattern, where two waves interfere with each other, occur when only one particle at a time goes through the slits? How does one particle, or one single wave, interfere with itself? Nobody has ever figured that out!

Strangest of all, when devices were constructed to detect which slit the particles passed through-- even when only looking at one of the slits, and not the other-- the interference pattern *disappeared* and the particles went back to behaving as particles would. Turn off the detector, and, voila! The interference pattern of waves re-emerged.

Bizarrely, the same phenomena have been observed when using decidedly un- wavelike molecules as the units; some quite large molecules, in fact. These shocking and incomprehensible results have been consistently repeated, thousands of times, in laboratories all across the globe.

All in all, the double-slit experiment was one of the great mysteries of science, still unexplained after long decades—centuries, actually-- of thought and analysis. And here was Merle, looking me in the face and telling me that I had a chance of understanding that phenomenal mystery! I think I was halfway in shock, just considering the possibility. I can totally understand why my dad was so excited to hear the explanation for such a long-held, tremendously mysterious experimental result.

Merle interrupted my daydream with a well-executed Earth gesture- a small fist bump on my left shoulder. And no gloves this time, either. "It's going to be quite a ride, Ken, if you're up for it."

The immensity of the mission was beginning to dawn on me now. The explanation of the double slit experiment- to be revealed! The true structure of the universe-- apparently quite different than what we now know, in some way-- also to be revealed! Even the realities of the strange enigma that is electromagnetic energy were to be revealed! The underlying reality beyond Relativity theory as we knew it-- somehow, amazingly, astoundingly, to be revealed!

I still remember the feeling I had right then, as I felt a recognition of the moment that was being thrust upon me. It was not unlike an electric shock—like my body was releasing massive amounts of adrenaline or something. I remember feeling my body actually buzzing, and I knew that I felt strong and ready, and full of an amazing energy. I was suddenly and almost overwhelmingly very grateful that Merle had chosen me. I gave Merle a firm little fist bump, right back, on his shoulder. "O.K., Merle. Let's do this thing."

Merle patted me on the back; in fact, he *thumped* me on the back, very enthusiastically, which surprised me just a little bit. "All right!" he said. "Let's do it, Ken! But first, let us take a stroll around the park, shall we?"

"O.K. Let's." And we both opened our doors and stepped out into the cool morning air.

# CHAPTER
# 16

A lot of people say that Merle must have planted these theories of physics and cosmology in my mind; otherwise, they say, I'd never be able to understand some of the so-called high-level physics concepts. I admit I had some of the same doubts, myself, and I came right out and asked Merle if he was planting thoughts in my head. He told me he was just allowing me to work it out for myself, as we talked about it, and I believe him.

Also, I was majoring in Physics at college, and I had been exposed to relativity theory, quantum mechanics, cosmology and all that, so I wasn't unfamiliar with the subject, by any means. Actually, I was particularly interested in the weirdness of Relativity, since high school, really.

Merle told me, right from the beginning, that he wasn't going to give me all the answers just like that. Not to say that Merle didn't put it right on the tee for me; of course he did. That's the primary reason he had traveled all those light years; to put that ball on the tee for me and ask me to take a whack at it. Of course, sometimes he had to hit the ball himself, when I overly struggled with my unlearning.

I believe that a big part of the reason Merle chose me, as opposed to a far more experienced professor of physics or something, is that my mind had not yet been hardened against other possibilities. I was still naïve enough to find much of Relativity, modern Cosmology and the Standard Model to be unreasonably strange and unexplainable. Instead of simply shrugging and accepting the weirdness as something no human could understand, I was willing-- with a great deal of encouragement-- to consider that there may be another angle to the whole thing.

In school, I really wanted to understand the space-time continuum, but I think I did have a subconscious nagging suspicion, due to the weirdness, that something very basic was missing from the story. It seemed like there might be one little loose thread that, when pulled, might reveal so much more underneath. There was always so much strangeness about the ratio of space to time, $c$. It was not only the ratio of space to time, but it was also light speed, no matter how fast the light source may be approaching or departing, somehow! It was also, incomprehensibly, the velocity limit of anything in the universe. Even stranger still, it was the limit set by the extremely bizarre relativistic addition of velocities, which I was already beginning to question again.

For that matter, the entire fabric of the Big Bang Theory seemed pretty far out there, with its counter-intuitive expansion, or Inflation, of space, and the mysterious beginning sequence. There was also the mystifying matter of what had occurred *prior to* the beginning sequence. And there were also some holes to be filled in on the Standard Model of particle physics, at the other end of the spectrum, with the little question of just exactly how fundamental particles acquire mass still looming large, notwithstanding any number of ongoing advancements in that field of study.

I knew that, for many physicists, a new breakthrough theory, or multiple breakthrough theories, had been widely expected for decades-- and arguably for centuries, in the case of wave/particle duality. For that reason, early on, I assumed that there would be an enthusiastic reception in the physics community, as word got out that I was offering strong and more sensible options to these previously confusing areas of theory. Merle tried to tell me that entrenched opinions are very difficult to reverse, even when using unassailable logic. He told me that any new information I had would never be enough proof for many people, and that many detractors would search out any small detail in which to derail my entire theory, in their minds, at least. I realized how true that was, not long after I first came out publicly with details of my Enlightening.

One of my neighbors had a degree in Physics, and he worked for an outfit that produced physics modeling software. So he was a professional in the field, and-- understandably, I suppose-- he considered me to be a misinformed goofball, operating beyond the fringe; just a kid who took a few physics classes and had a dumb idea that didn't even make sense. I remember once, when I encountered him on the sidewalk, walking his dog, he told me that anything I said about Hyper Relativity was completely useless without me conducting repeatable, physical experiments to support my theses.

I tried to tell him that describing the hyper-dimensional universe is theoretical physics, like Albert Einstein's Relativity theories were; done as thought experiments, perhaps, or as mathematical equations, but not necessarily as actual physical experiments. I tried to tell him that many experiments pertinent to Hyper Relativity, such as the double slit experiment, or the Michelson- Morley experiment, had already been performed, by others. But before I could get more than a sentence or two out, he grabbed hold of his dog's leash and practically sprinted away from me, half-dragging the dog along, in order to avoid having to listen to my side of the story.

I realized that he probably didn't even understand what I was talking about, anyhow, since he was involved in a more specific type of physics work; sort of a niche knowledge. But nonetheless, he sure seemed to know that I must have it all wrong, because I didn't peer through a telescope, or measure light "waves", as had already been done so many thousands of times before, to no avail!

In fairness to me, though, this book does contain a generalized description of one pertinent experiment that I helped Merle perform, in an actual laboratory here on Earth, that totally clarified the reality of "wave particle duality" for me. That's the experiment that made milk come out of my father's nose, when I described it to him.

Anyhow, since that "conversation" with my neighbor, I have repeatedly found that it is true, with a few notable exceptions; people do not change their minds very easily. For me, however, most of my newfound scientific knowledge seemed to be rather straightforward. Each revelation seemed so simple and obvious, in retrospect. Once I saw it, my mind would boggle as to how it had gone unnoticed or unknown for so long.

And each simple and obvious truth led astoundingly to another, greater revelation, but equally simple and obvious, in retrospect. And so on. It's really not difficult for me to believe Merle's point, that it was unusual, and most likely detrimental, that we had come so far in our science while remaining blind to the hyper-dimensional universe, and the hyper-dimensional greater universe.

As far as mathematical formulations, the equations necessary to understand Hyper Relativity as a concept already exist, in the form of existing relativity theory. What some people don't realize about Hyper Relativity is that it requires the *removal* of a mathematical equation from Special Relativity, as opposed to the *addition* of an equation or equations, as well as a simple revision to a postulate and maybe a little rethinking here and there. It does not require a full-blown demolition of the original theory. It does beg a great deal of additional mathematical exploration, of course, as any new thinking in Physics must, virtually by definition. A vast new field of multi-dimensional mathematics will undoubtedly open up, as the new possibilities of the hyper-dimensional universe become clear to us.

Nobody ever realized that Albert Einstein vastly limited his own theory by trying to take it too far, in a sense. Now I know that Einstein never truly could have understood the actual physical mechanisms of the universe that would result in some of the strange parameters of existence as he understood them to be, due to his misinterpretations of the physical reality that the Lorentz transformation equations actually represented.

To me, Hyper Relativity is really a simplification of Relativity, since it describes clear physical mechanisms for the strangeness of our interactions with the space/time continuum, throughout the infinite and eternal universe, and, truly, even *beyond* our own universe. It amused me to think that Buzz Lightyear's famous line might not be as contradictory as it sounded.

As far as additional new experiments of verification, it seems safe to say that there will be a vast bounty of physical experiments and observations to come, both in cosmology and particle physics, in particular. These new equations, and these new experiments, are beyond my own resources or abilities. However, the experiment that Merle and I performed, which I describe in this book, is actually a quite simple type of experiment that can be approximated at an appropriate facility with existing off-the-shelf technology, and probably at a modest cost, once the specifics are worked out.

Any limitations of mine, in terms of financial and experimental resources, do not overturn the basic premises of my unlearning, however. I now understand the meaning of the equations of Hyper Relativity at a level far beyond what I had previously understood in conventional, modern Relativity theory. My eyes have been opened to a supremely greater universe, by casting aside the hobbling limitations placed on the original Theory of Special Relativity. Because I was willing to take a big step back, and unlearn that which I had already learned, I was able to take some giant leaps forward in my own knowledge, thanks of course to Merle's great patience and wise guidance.

# CHAPTER
# 17

Merle's love of the people and music and general culture of Earth seemed only to be eclipsed by his love of the natural world. We strolled into the park, and we hadn't gone more than ten or twenty yards before Merle noticed a cicada, hanging onto the bottom side of the branch of the tree it had just hatched out on. Its still-damp wings were hanging straight down, glittering like iridescent green emeralds in the oblique, warming rays of the early morning sun.

"A cicada," I said.

Merle strode up, peered closely at the cicada and announced its scientific name, as he usually did when he scrutinized some bug or tree or whatever. "*Tibicen canicularis*," he said. I never saw a bug that stumped Merle, except one time there was some small little fly of some kind that he actually didn't recognize. But the next day, he told me the scientific name of the fly. He said it was unusual to see that species so far north.

Anyhow, that cicada and its discarded nymphal exoskeleton fascinated Merle for several minutes. He intently watched the insect, from probably six or eight inches away, as it hung there waiting for its wings to harden. Merle circled all around to view the various body parts. I remember that he was talking about how the male cicada makes its loud sounds, by alternately clicking the tymbals on either side of its body, and using part of its body to amplify the sound.

At one point, I started to reach out towards the cicada, to touch it, and Merle grabbed my wrist. A cicada is very vulnerable at that point, newly hatched out, with wings still wet, and I think Merle thought I was going to grab it or something. Actually, I was just going to touch its back—not the wings. In no way was I going to try and grab that thing, even though a cicada is actually completely harmless to people. A tree might have a minor complaint, though.

I often got the impression that Merle was documenting his entire visit by scanning or recording what was happening. Whether it was through his eyes, or maybe a hidden camera or cameras, I don't know. Although he did it in a way that was not very obvious, I came to notice that he was always looking around, in a full circle if possible. I also notice that he looked up, into the sky, quite frequently, and I asked him once why he looked up into the sky. I had been thinking that maybe Merle was looking for spaceships up there.

Instead, Merle seemed surprised that I even asked the question. "Why do I look up into the sky?" Merle asked, repeating my question. "To see what might be up there, of course. Around here, you might see a butterfly, or maybe a cicada bumbling along between the treetops. Innumerable beetles and flies and bees and other insects are up there, also, which I find interesting.

"You can always check on the weather, by looking at the sky. Or you might just enjoy all the different cloud formations because they can be so interesting in their patterns, and often very beautiful, especially early or late in the day. Then again, a crystal blue sky is beautiful, as well. We have a saying on Akeethera that 'nature creates the finest art,' and I do believe that is true.

"Another nice thing is that your moon may be visible at any hour of the day or night- as it is right now, in fact." Merle pointed upwards at the pale and unobtrusive daylight moon, high in the sky. "I love your moon because it is so close and large. Akeethera has three moons, but they're each much smaller than your single moon is, and you can barely see them in the daytime. Your moon is so interesting to look at. And I am fascinated by its history." Merle peered up again. A passenger jet was passing along through the sky, nearly passing in front of the moon.

"There are almost always airplanes up there, as well," he noted. "I'm very interested in all the different types of planes that are flown around here. There are two or three international airports, plus several smaller airports, relatively nearby. So you get to see a lot of interesting airplanes in the air, large and small. Sometimes there are helium balloons, or blimps, up there, also. Or drones, or kites, or birds, or many other things." He pointed to a patch of sky, and took a deep sigh. "Also, Akeethera is up there…"

Here, Merle was trying to confide in me about how he missed his home, I believe, and I just blunderingly interrupted him to ask the question. "Merle, can you see spacecraft up there? Or are you looking for spacecraft?"

Merle sort of smirked a bit at that question. "You're not going to see very many uncloaked ships out there," he said. "When you saw my base ship, they purposely de- cloaked it, just for your benefit."

"Oh." I said.

"However, if you *did* look up into the sky on a regular basis, you'd probably have some chance to eventually see a ship, someday. It's possible to occasionally see a de- cloaked ship, for all kinds of reasons, such as technical difficulties, or, occasionally, a deliberate visual manifestation. Certain types of probes or ships aren't really even fully cloaked, and those can be visible. Usually those are unconventional-type vehicles that result in fairly inconclusive sightings, though, like balls of light, or whatever. But if you set out looking for a ship, or a probe, it'd be difficult to get that lucky, on any given day. So I can definitely say that I'm not looking for spaceships, specifically. But it would be very cool to see one, other than my own, definitely." Merle grinned.

I imagine that Merle also kept his head on a swivel just for purposes of safety, like a fighter pilot, especially after what nearly happened to Magu. For all of Merle's enthusiasm, he must have experienced a certain measure of anxiety, as well, venturing about on a dangerous planet such as Earth. Merle once told me that Akeethera had been monitoring Earth for several thousand years, but that nobody from his planet had ever spent as much time on-planet at one time as he and his fellow explorers were spending.

"You would be surprised," he said, "to learn how similar Akeethera and Earth are to each other, in many ways. They are sister planets, you might say, in more ways than you might think." Then he gave me a wink, as he liked to do when he closed the door on a subject. That was all I could get out of him on that subject of "sister planets," although in the years since, I have pondered that subject many, many times.

Anyhow, we hadn't gone more than another twenty-five yards or so before Merle crouched and peered down at the edge of the pathway. Suddenly he dropped to his hands and knees on the path, to more closely examine a large cricket that I imagined was probably heading back to his cool daytime nook, to avoid the blasting heat of the sun and sleep the day away, after a long night's adventures.

"*Gryllus veletis*!" I swear Merle practically squealed when he sighted it. He'd always get so excited about that sort of thing. He turned to me for a moment and told me the common name, "It's a Spring Field Cricket". This one Merle picked up, in his uncannily quick and smooth method, but he immediately set his hand on the sidewalk and opened it, allowing the insect to hop back out onto the sidewalk, unscathed. As the cricket hopped into the crack at the edge of the path, Merle mentioned that the cricket makes a sound in a completely different manner than the cicadas do; the cricket does it by rubbing its wings together. I think I had heard that before somewhere, how the cricket uses his wings to make the sound. I always figured that's what a cicada was doing, also, but the cicada's system is a lot different, actually. I guess that's why it's so crazy loud, in the summer, if there are just three or four cicadas buzzing in the trees.

I must say it was oddly exciting to be there as Merle checked out every little mundane creature or object he came across, with the barely-contained enthusiasm of someone who was fulfilling a very long-held and precious dream. Merle seemed to consider every milkweed pod, squirrel, oak tree or blue jay to be amazing and priceless. As I observed Merle's almost fawning reverence for the local flora and fauna, I found myself thinking about Merle's home planet Akeethera, and what type of insects and animals roamed its plains and forests. I wondered if I would have had the nerve to reach down and pick up an alien bug, like Merle did with the cricket. I suppose I could probably do it, especially with years of training.

I do think that witnessing the joy Merle experienced through those interactions had a great and positive impact on my life. I remember that I once mentioned to Merle that I enjoyed the tremendous gusto with which he approached each bug, tree, or sunset. I told him that he probably even took pleasure in a traffic jam.

"The universe makes very few guarantees of the future, Ken," he told me. "It *does* guarantee that there will be a tomorrow, but it can't guarantee that *I* will have a tomorrow. Not to mention that even if I am still here tomorrow, I may not be able to see or do the things that I am able to see or do today. So I want to appreciate all the great gifts there are to appreciate, today. That traffic jam might seem inconvenient, but for most people there will come a day where they are physically unable to sit in that traffic jam. Many, many, people already find themselves in that situation. And many of those people would sacrifice almost anything, to be able to have the mobility to get out there on the highway, and fight the good fight, and battle through a traffic jam, again." He thought about it a little more, in silence, for a few seconds.

"Also, there's usually something of value to be learned, in just about everything."

"So what's to be learned about a traffic jam?"

Merle thought about that one for a moment. "So many things can be learned, out in traffic. I especially enjoy being out on the four-lane highway, during rush hour." Merle gestured in the direction of the expressway, which was a mile or two away from us.

"During rush hour? It's terrible out there then, isn't it?" I couldn't believe Merle said he actually enjoyed that.

"Yes, the traffic is terrible; just a crawl. But I enjoy the human interactions out there. It's quite an amazing experience, really. I've never experienced anything else quite like it."

"How is it amazing?"

"Oh, in so many ways. For example, why do cars have horns?"

"To warn other drivers, I guess, when something dangerous is happening."

"Is that why people use them so much in traffic jams?"

"I guess not, really. They use them more to complain about other drivers, probably."

"That's what I've noticed, also. They use their horns to chastise other drivers, mostly, in a traffic jam, even though nobody can really go anywhere, anyhow, in many cases."

"I guess that's true."

"But you see some friendly and kind gestures by drivers, also-- more noticeably so in a traffic jam than when traffic is moving well," Merle said. "There is also one particular act of courtesy that I see which actually makes these traffic jams worse."

"Makes them worse? What do you mean?"

"When one lane of the highway is going to be closed up ahead, it seems like everybody is in a big hurry to get out of that lane, long before the actual merging point. That instantly causes a traffic jam in the remaining lanes. Meanwhile, the lane that is going to be closed up ahead is often left wide open, with no traffic whatsoever! If a car tries to use that lane, which is the smart thing to do, many of the other drivers actually seem to be angry about it! I've seen semi-tractor trailer operators purposely move partially into that lane, so nobody can use it, as if they are the arbiters of highway justice! On occasion, I've seen traffic crawl for over a mile, while all the while there is a completely open lane that nobody is using! It's the most confounding, ironic thing I think I've ever seen!"

"Ironic?"

"Yes, of course! So many drivers are rude to others out on the highway, which causes traffic slowdowns all over the place. But the one area where many of these same rude drivers appear to have a sense of justice is in this merging situation, where their rare cooperative intent is truly misplaced. Finally, there is some civility, but it's counterproductive, and actually turns a minor traffic inconvenience into a gut-wrenching jam-up, and leads to more angry drivers! I sent video of the phenomenon to some of my friends back on Akeethera, and at first they thought it wasn't even real—they thought I faked the video as a joke. Of all the videos I sent, that one clip got the most interest from people, by far. A lot of people just couldn't understand the reasoning behind it, even though I told them that the drivers here were just trying to be polite."

"So what are people supposed to do in that situation?"

"If you are driving in the lane that is closing, don't leave too soon! Use the open highway while it is there! When the lane finally does close, then two lanes should merge like a zipper, with alternating cars from each lane moving ahead. Simple!"

I was actually amused by Merle's near-agitation over this subject of the "zipper merge." I wanted to needle him a bit, by telling him I thought he said he *liked* traffic jams, but I thought twice about it, since he did seem just a touch agitated, as I mentioned. Instead, I thought of what traffic must be like on Merle's own home planet. "Merle, are there traffic jams on Akeethera?"

"No, not really; not like here, anyway. We do have some traffic, though."

"Do your cars roll on the ground?"

Merle laughed at that. "No, they don't," he said. "They go pretty fast, too."

Merle didn't want to tell me much more about how they get around on Akeethera. So I asked him why our jams are so terrible, other than the zipper merge issue.

Merle looked at me, semi-incredulous at the question, I think, as if he couldn't believe I might not know the answer, myself. "Well, a lot of the time it's as simple as somebody pulled over on the shoulder and getting a traffic citation, which makes people slow down. If just one driver slows down to look, the next hundred cars behind them also have to slow down, to avoid crashing. So if just one driver out of every hundred slows down to gawk, it's nearly full gridlock.

"Or maybe there was an accident blocking one of the lanes, or road repair, or something, which always slows things down. And then, of course, there's the downlookers."

"Downlookers?"

Merle pantomimed somebody looking down at their cell phone as they were driving. "Oh, right," I said. "Downlookers."

"They can be a little disengaged," Merle said. "Primarily, though, day in and day out, it's that people don't seem to know how to use the passing lanes," he said.

At first I thought I misheard him. "The passing lanes?"

"Yes, the passing lanes. So many people don't follow the basic rules of the passing lanes! Slower traffic should be in the right lanes, in your American traffic system, but a lot of drivers seem to just pick a lane, and stay in it, at their own speed, regardless of the circumstance! They should be watching for other drivers coming up behind them, in their rear-view mirrors, and they should move to the right to let the other driver pass, but many just stay in their own lane, no matter what. They apparently don't look, and if they do, they don't seem to care, for whatever reason. The result is clusters of tightly- bunched vehicles, with cars in the back blocked from getting through, that are interspersed with huge open spaces of highway. It's those huge open spaces that cause the backups, the same as with the wasted open highway in the zipper merge situation.

"In fact, I have observed that some drivers seem to actually enjoy blocking other drivers, purposely, by impeding the passing lanes. I suppose they imagine that they are gaining an advantage, by keeping everybody else behind them. Thank goodness everybody doesn't employ that strategy! Traffic would grind to a standstill, very quickly!

"A lot of the passing lane violators, I believe, are simply oblivious to other drivers. They possibly couldn't care less whether they may be blocking a passing lane, and inconveniencing another driver. Also, it's likely that many drivers aren't even clear on the concept, or are not skilled enough drivers to change lanes properly, as is typically required in a passing-lane type traffic system. In that case, they should probably stay in the right lane, instead of one of the left lanes, but I imagine that many aren't even aware of the protocol.

"Then, of course, there are the drivers who tailgate, and the traffic weavers. A lot of the time, they don't get anywhere too much faster, anyhow, but their erratic driving causes others to slow down to avoid an accident—if they can avoid the accident.

"Fortunately, there are enough people out there who do look out for other drivers, and follow the passing lane rules, and allow merges without battles, and that sort of thing, so that the traffic still manages to lurch along, somewhat. All in all, it's quite an incredible scene!"

I was tremendously amused by Merle's heartfelt description of an Earthly traffic jam.

In the years since, while crawling through traffic, I've often thought about Merle's comments. It's interesting to look at traffic jams in terms of human interaction, instead of just too many cars, like I'd usually thought of them in the past. In some ways, a traffic jam is like a pantomime of human existence, punctuated by honks and thrusting fingers and questionable decisions. Occasionally, some choice words get shouted out of an open window. On the other hand, you'll see acts of kindness and grace, as Merle had mentioned, and the occasional "thank you" wave of a hand, right smack in the middle of the most frustrating jam-up!

Sometimes I think we'd hardly ever have traffic jams, if people could just get along better. All in all, though, I still say it was easy for Merle to say that a traffic jam can be enjoyable. I'll bet that if he sat through as many as I have, he might think about it a little differently!

Merle continued speaking. "Also, Ken, I am painfully aware that my time on this planet is short. Any traffic jam that I am in may, indeed, be my final traffic jam. Any cicada, or cricket, that I might see, might be the last one that I ever see. The same is true for all people who walk this Earth, and all who walk *any* world. As I said before, who knows what tomorrow may bring?"

Merle often referred to commonplace things that we so often take for granted, or even complain about, as "treasures", or "great gifts." That was probably some of the most important enlightening I got from my time with him. I am always struggling to keep in mind, as I go through life, to not take things for granted. It's not easy sometimes. But you can find a lot of happiness in the simplest little things, if you realize what a gift they truly are.

We continued to walk along, until we came upon a park bench. The park was still largely empty of people at that still-early hour, but as we approached the bench, a father and his young son rode up on their bicycles, and they began unloading a bright red model rocket, and the launching pad with other launching supplies, out on the lawn area. Clearly, this less-crowded early morning hour in the park was a good time for a model rocket launch. Merle was in his glory, in his observation mode, and we sat down to watch the launch.

After they had set everything set up, the boy crouched down to press the button to initiate the launch sequence. The rocket erupted off the pad and blasted off into the sky with a swirling trail of white smoke, and Merle responded by loudly clapping and cheering for the pair of rocketeers. "Bravo! Bravo! Excellent!"

The boy, probably about nine years old or so, hopped about enthusiastically, high- fived his dad over the perfect launch, and shot an approving glance back at Merle. The young boy looked up with uninhibited joy as the billowing, bright yellow parachute opened on cue, far above, and a bit to the southwest, in the endless blue summer sky. The boy's father gathered up the launching pad gear in a bag, and they jumped on their bicycles and gave chase across the park, trying to beat the rocket to the landing spot. We watched them recede in the distance, bouncing as they high-tailed it across the grass of the park. It looked like they perfectly planned the trajectory, taking into account the direction of the light morning breeze out of the southwest, so the rocket would land within the park confines, out in the large central clearing area, away from the trees.

"Nicely played," Merle observed.

# CHAPTER 18

Merle leaned back in his seat, put his hands behind his head with elbows out, and extended his legs in a relaxed posture. He emitted a long, deep sigh. Then he asked a simple little question.

"Ken, who do you think was the smartest man that ever lived on Earth?"

"Oh, I don't know. Probably Albert Einstein, I guess."

"What about Sir Isaac Newton, or maybe Leonardo da Vinci?"

"Well, I don't-- I mean, yeah, I guess they were really geniuses too--"

"Oh, I'm just having a little fun, Ken. I think most people these days would probably say Einstein, wouldn't they?"

"Sure."

"But at the same time, you have questioned some aspects of relativity, haven't you?"

"I guess so."

"So do you think Einstein may have been wrong, in some respects?"

"Oh, no, I don't think so, any more. Or, at least, I *didn't* think so. I thought that relativity has been very well proven by experiment. But maybe not, right?"

"Well, *portions* of relativity have been very well proven by experiment, like time dilation, and gravitational waves, for example. But what if *other* portions are not quite right? Do you think people think about General Relativity or Special Relativity critically, or do they just unwaveringly follow along the trail of full and complete acceptance?"

Well, that was a big moment of impact for me. I realized what Merle was essentially saying. Nobody learns about the Theory of Special Relativity, for example, without completely and utterly accepting that the two "Postulates" are true, and without accepting that the weirdness of the whole $c$ thing is completely true, even though these ideas defy the parameters of physical interaction, as far as I could ever tell.

Oh, you might dig in your heels a bit, like I did, questioning, at first. In the end, though, you just accept that certain aspects of the universe, according to Special Relativity in particular, seem to have plunged down the rabbit hole, and through the looking glass. You have to be willing to suspend your disbelief, say to yourself "that's just how the universe is," and move on to the next subject. So it wasn't hard to think that the entire mountain of relativity, in academia, is based entirely on the complete acceptance that all of Albert Einstein's basic stances in terms of General Relativity and Special Relativity were all totally on the money.

"It is a very deep dogma," said Merle. "It's a very tough nut to crack."

"Yes," I said. I think at this point, I was still just trying to wrap my head around the thought of relativity theory being off-base, in some fundamental manner, apparently, even though I had already discussed the Lorentz transformations with Merle on a couple of occasions. I just couldn't begin to imagine what sort of universe that may imply.

Merle continued. "But yet, we already know that Einstein wasn't perfect. His vacillations over the cosmological constant, for example, are well documented. So perhaps there is some hope that people may be willing to consider another possibility."

I nodded, and we both thought in silence for a few moments, before Merle continued.

"I'd like you to do me a favor, Ken. Let's just look, specifically, at Special Relativity, right now. What would be some aspect of Special Relativity that you find to be especially strange, and difficult-- if not impossible-- to understand by conventional physical means?"

"I don't know. A lot of it is weird, I guess."

"How about the mass-energy equivalence equation?"

"You mean, 'E equals mc squared'?"

"Yes."

"Well, to me, that equation means that all energy is convertible to mass, and vice versa. Any object in the universe, with mass, is basically a conglomeration of energy, and can be converted into various forms of energy. Or, the mass of an object is directly analogous to the energy of an object."

"That's right. Basically, mass and energy are converted forms of the same thing.

That's why I like to use the term 'mass-energy'. It's a more inclusive term."

"Yes, that seems to make sense, actually."

"Okay," said Merle. "That's good, Ken! Now we're getting somewhere! This is how a detective might approach it, Ken. Go through the list of what we know, and look for oddities-- the anomalies that may seem hard to understand. If it seems fine, like this equation, then move on. Maybe we might need to go back later, for another look, but as for now, we'll move on."

"Sure." I said.

"How about the first postulate, then?" Albert Einstein, in his Theory of Special Relativity, included a "first" and "second" postulate, as two prelude cornerstones to the theory.

"So what was the first postulate, again?"

"Basically, that all the laws of the universe are the same between all uniformly moving frames of reference."

I liked to put things in terms of specific examples, sometimes, for clarity, so I came up with a good example to illustrate the postulate. "So in other words, if I took off from here at 80% the speed of light, and you flew right alongside me, also at 80% the speed of light, you and I would experience all the laws of the universe in the same way, since we've both accelerated to the same velocity."

"That's right."

"Well, that sounds about right. Two people sharing a common space/time reference frame within the space-time continuum."

"As you and I are, right now, just standing here."

"Right. So in my example, another person who stayed behind, and didn't fly off at 80% the speed of light, would look at and see things differently, since he or she would be in a different frame of reference, velocity-wise."

"That's right."

"Well, yeah. That seems good. The first postulate actually seems to make normal enough sense, also. I mean, that's how I picture the space-time continuum to work."

Merle sat up forward in his seat. "Good! Good, Ken!"

Now Merle raised an eyebrow and asked the follow up question, which he no doubt had been looking forward to asking me, for some time now. "Now what about the second postulate?"

"Well," I said, "that one is a bit weird." We had already discussed the second postulate, previously, where the speed of light always appears to be the same, to all observers. "I guess I never really did understand how the second postulate works," I conceded.

"Maybe that is a clue."

"A clue?"

"That's right. Maybe the weirdness of light speed is a clue to something about the universe."

"Like what?"

"Well, what is involved, when electromagnetic radiation travels?"

"I don't know. A guy and a flashlight, or something?"

"No, Ken. That's not what I meant. Maybe I should have asked what *two things* are involved, when electromagnetic radiation travels?"

"Two things?"

"That's right. Obviously, one of the things is the electromagnetic radiation, itself, in the form of photons, which are discreet units of energy, or quantum mass-energy, if you prefer."

92 | *The Enlightening*

He had me stumped with the second thing, and I sat there, trying to think, for several moments. "I have no idea, Merle."

"You mentioned it when we talked about the first postulate."

"When we talked about the first postulate... hmmm... Oh! The space/time continuum! The answer is electromagnetic radiation, and the space/time continuum!"

"That's right."

"So what is the clue, though, exactly?"

"Perhaps the strangeness of light reveals something fundamental about something other than light itself."

"Something other than light... Oh! It's revealing something fundamental about the space/time continuum!"

"That's right. Also, there just might be some interesting little details involved in the physical mechanism of electromagnetic energy, itself, also."

I couldn't really process much more, I don't think, at that moment. I did have a good clue to work with, though. The $c$ weirdness of light has something to do with the nature of the space/time continuum itself, in addition to some kind of physical trick involving the mechanism of the electromagnetic energy itself. "Let's mark that down as an oddity, then, Merle."

"Yes!" Merle said. "An oddity it is." Merle grinned at that. "And what else about the theory seems a little hard to understand?"

"Probably just the whole thing with the Lorentz transformations, like we've already talked about, and with $c$ being the maximum speed limit of the universe, and the relativistic addition of velocities, and all that."

Merle smiled. "It is possible that the real-life significance of a mathematical equation, or a series of equations, might be misinterpreted, isn't it?"

"I guess so."

"Remember," Merle said, "the reason that the Relativistic Addition of Velocities exists, in part, is to rationalize the speed limit of $c$ in the universe which Einstein had postulated, which itself was based on his misinterpretation of what the transformation equations signify, in our physical universe. Also, in fairness, Einstein had no compelling reason to consider that the universe consists of multiple dimensions, or completely separated reference frames of space/time. The equation also ties into the concept of the second postulate. The Relativistic Addition of Velocities was based on the first and fourth equations of the Galilei transformation, really, and folding that into the first and fourth equations of each Lorentz transformation.

"Einstein basically stated that the Galilei transformations do not hold *in reality*, while he maintained that the Lorentz transformations *do*, indeed, reflect reality. I would argue that a true understanding of reality requires the balanced understanding of what both the Galilei and Lorentz transformations indicate about reality, and our perceptions of it.

In a way, the Lorentz transformations reveal how the universe *disguises* the truth. But in another sense, they reveal a greater truth, if we allow it to."

Merle was starting to lose me, although I was genuinely trying to follow him.

Merle continued. "So Einstein, in his theory, uses the Lorentz transformations, which really only involve perceptual relativistic artifacts between an observer and a traveler, and he formulates a comprehensive, universal law, describing these artifacts of perception as the actual full reality of the entire universe, resulting in this Relativistic Addition of Velocities." Merle frowned, which was the first time I had seen him frown.

My coy alien friend had to stay within his boundaries, I suppose, but still, he must have been itching to just spell the whole thing out to me, all at once. Merle was more patient about it than I would have been, I think. Nonetheless, at this point, I was starting to look at the first few pieces of the puzzle a lot more seriously than I had previously.

At this point, I became aware of a group of three women approaching us from behind. It was the time-savers! I was feeling much friendlier towards them at that point, knowing some of their incredible background. I was looking forward to talking to them some more. I turned and said "Hi". I probably did a bit of a double when I saw Clotro. She looked absolutely amazing. It was the first time I had seen her with her white hood down, and her long blonde hair shimmered in the bright sunlight like rippling waves of gold. She seemed to be enjoying the day as much as Merle was. She was definitely turning a few heads from some of the passers-by. She gave me a nice smile and nodded slightly. Latsis dipped her head ever so slightly as an acknowledgement. Atropha, however, looked like she was in a bad mood again.

"What is the status?" she asked Merle, rather curtly, in English.

"We were just talking about the Relativistic Addition of Velocities equation from their Special Relativity Theory," Merle told her.

"Oh," said Atropha. "That. And what does Ken think about it?"

"We haven't gotten into it very far, yet."

By now Atropha was standing in front of us, with the two others off to the side. She looked directly at me, in a purposeful and somewhat exaggerated manner, with those penetratingly deep black eyes. "Well, Ken. See to it that you keep an open mind. In the end, you'll have to decide for yourself." And with that, she spun around and gave a "follow me" wave to the other two. I noticed that she also signaled to Clotro to put her hood back up, which she did immediately. They strolled away from us, side by side, back behind a dense cluster of bushes. I never saw them come out from behind the cluster, so I assumed they had gone back to the ship in their usual manner.

"Now, where were we?" Merle asked.

"Well, Merle," I said, "the Lorentz transformation equations must be correct, mathematically, while our interpretation of what they signify in the universe is not quite correct. And the Relativistic Addition of Velocities is just a mistake in general, in some way. And light speed is really some basic feature of the space/time continuum, and maybe something in the way in which light itself travels has something to do with it, also. Oh, and the speed limit of $c$ is-- maybe it isn't right. Oh wow, does that mean you can go as fast as you want?!" The thought of going faster than light was very appealing to me!

Merle looked at me, straight-faced, for a good long moment, before he broke out into a wide smile. "Well, Ken, I already told you that I myself traveled faster than light speed, or, more accurately, the ratio of space to time, on the way here. But even if you might be correct in what you say, you still don't understand the context, or the physical mechanisms involved, or any of that. When you are making adjustments to the Theory of Special Relativity, then surely there must be other adjustments to be made, in how you think about the universe."

"True."

Merle leaned forward in his seat, his right hand tucked under his chin, as he gave the matter some more thought. Then he sat back up straight, and turned to face me as he continued speaking. "You know what, Ken? We should take this back a few steps, back to the basics."

"O.K., whatever we need to do."

"There are some basic concepts that all young students on my home planet, Akeethera, learn, like how to write, how to read, basic math, and all that. But one of the first things they study is perhaps the most basic question there is."

"And what is that?"

"The question is, 'what two things comprise the universe?'"

"What two things comprise the universe?"

"That's right."

"And what's the answer?"

"What do *you* think the answer is?"

"Uh, matter and energy, I guess. Right?"

"No. You are almost half-way to your answer, though."

"I'm almost half-way to my answer?" I had to think about that for a bit. "Is it mass- energy?"

"Well, *now* you're half way there. Mass-energy is *one* of the things that comprise the universe." He waved his arm around, slowly, in an all-encompassing manner. "Mass- energy represents most of what we visualize when we speak of 'the universe'. You and I are mass-energy. Earth is mass-energy. The Milky-Way is mass-energy. The movements of your arm, or of the stars, are mass-energy. A burning flame, the dance of a ballerina, the flight of a damselfly-- all are mass-energy.

"Mass-energy, Ken, is an entirely quantum phenomenon. Every form of mass or energy, anywhere in the universe, exists or originated in discreet, separately formed units of energy, such as individual sub-atomic units, or individual photons. That is what is meant by quantum—formed of individual, discreetly separate units."

I interrupted Merle. "But what of the wave-like nature of particles? Isn't that a sort of non-quantum behavior?" Well, you would think I struck Merle with an electric shock, the way he bolted in his seat when I said that.

Merle literally shouted to me, as he slapped the seat next to him. "Absolutely! *Absolutely* it represents non-quantum behavior!" He leaned in closer to me, and looked around, as if he just realized that he had been shouting. Then he asked me a question, in a much lower, even hushed manner. "The thing is, Ken: Is that non-quantum, wave-like behavior a characteristic of the particle itself? Or might it be a characteristic of the second thing that comprises the universe?"

"The second thing, right?"

"Yes. But what do you think it is?"

"The space/time continuum?"

Again Merle jumped as if he had been shocked. "Yes! Yes, Ken, yes. The space/time continuum is the other thing that comprises the universe! The space-time continuum, you see, is completely non-quantum. That's why we call it a continuum, of course. A continuum, by its very definition, is not in any way divisible into discreet, individual units. It is totally non-quantum, yet the continuum itself *does* represent a form of energy, that interacts with our quantum world of mass-energy. Can you guess how it might interact?"

I had an idea, based on what Merle had been saying. "The space/time continuum interacts with mass-energy in the form of a wave?"

"Yes!" Merle looked relieved. "The quantum particles of mass-energy ride the non- quantum wave of the space-time continuum." He paused and looked up into the sky for a few moments, watching a large flock of starlings fly past. "We have a saying on Akeethera, Ken. Roughly translated to your language, it goes 'we are all but ducks on a lake'. The ducks represent mass-energy. The lake represents the space-time continuum. It's not a perfect analogy, but it's not a bad one, either."

"You mean that you have ducks on Akeethera?"

"Well, we have flying creatures that are partially aquatic, much like ducks. So it's the same idea, and I'm just using 'ducks' as the nearest translation."

Merle's analogy seemed pretty simple to understand. "Merle, that is like the secret of the universe, right there! That solves a huge, central question of quantum mechanics, right there!"

Merle smiled at that. "Well, I told you that the universe doesn't hold any secrets, Ken. And don't get too excited, just yet. You still can't put that into much of a context. You still need to fill out your understanding of the entire situation. And you need to understand the space-time continuum to a much greater extent than you do now."

"I think I understand space-time pretty well, now, Merle."

"You do? Well, a positive attitude always helps, doesn't it? We will talk quite a bit more about the space-time continuum!" At that point, Merle stood up from his seat on the bench and took a few moments to slowly stretch his body, this way and that way, like a cat does after a nap. As he stretched, he continued conversing. "We should let these ideas we've bounced around soak in for a while, Ken. We can continue our Special Relativity talk later. In the meantime, let's keep walking. It's too nice of a day to just sit here talking, anyway."

I stood up and stretched, myself, actually relieved in some way to take a break from our physics discussion, in spite of being very interested as to where it all was leading. "O.K., Merle. You lead the way."

# CHAPTER 19

We left the path, and strolled across the large central grass area of the park. Stands of trees, shrubs, picnic tables, picnic shelters, and outhouses were interspersed here and there throughout the park. Merle, as always, was checking out every bug or bird he saw. Occasionally, random things like a drinking fountain or an outhouse caught his interest.

One thing that sort of took me by surprise at first was that Merle often reached down to pick trash off the ground. He collected it to deposit in the next garbage can we came across, so our walk was basically a zig-zag route between pieces of trash, and garbage cans, and we stopped at interesting sidelights along the way. I couldn't help but join in on the trash collection, at some point. Just the two of us picked up enough scattered refuse to fill up an entire park garbage can, at least, that day.

Eventually, we closed in on the northeastern portion of the park. Streamside Creek ran along a portion of the western edge of the park, as well as the entire northern edge. The baseball fields were in the northwestern corner near the stream, and the basketball courts were to the east, also near the stream and not far from the parking lot that ran along part of the eastern edge of the park. There were a couple of guys out on the basketball court already, and one of them was pretty tall. I knew who that guy was.

Tommy Marnel was, basically, the king of the basketball courts out at the park. In fact, to this day he still goes out there from time to time, maintaining his honorary role on the courts. He just can't spend as much time out there, these days, with a wife and a young daughter. He works at his uncle's factory, as a Production Supervisor, and he coaches his nephew's travel basketball team, too, so he definitely keeps busy.

Tommy's other uncle, George, was the varsity basketball head for our local high school, Rundle Central. He'd been the head coach for about eight years or so, by the time I tried out for team. I heard that he's coaching at some other school now; I forget where.

Both Tommy and his uncle George were big basketball stars at R. Central, back in their respective days. They both have plaques on the big wall of the Rundle Central Wildcat Hall of Fame, in the foyer of the Fieldhouse.

I played basketball at R. Central, myself, freshman and sophomore seasons. Back then, I was playing summer league basketball four days a week, for most of the summer, which was part of the basketball program. On off days, a lot of guys from the school teams would come out to Streamside, to try and get in there to play against Tommy, who was playing in college at the time. I played against him a few times and actually did pretty well, I thought. The last time I came out here, it was late July, before my junior year of high school, and I remember it was real hot out there that day.

Probably less than three minutes into the game, I was battling Tommy for a rebound, and he jumped for the ball too soon. As he was coming back down, and I was going up, his elbow came right down onto my nose, and my nose broke.

My mom freaked out when she found out I was hammered by Tommy Marnel, who she knew from when my brother was in high school. She said he probably outweighed me by 25 pounds. It had to be at least 50 pounds, but I didn't tell my mom that!

Mom never did like when I went to play hoops at Streamside Park, because of some of the "unsavory characters" who sometimes hung out there by the courts, and in the parking lot. So, after my nose broke, my mom made me promise not to go out to that park to play against any college kids until maybe the following summer. I battled her on the issue, but in the end I had to promise to not go back there. For the rest of that summer, I played basketball at Reclamation Park, closer to our house.

That fall, somebody at school told me that someone had asked about me, out at Streamside, and Tommy said that I was afraid to come back, because of the broken nose, and because of how rough it was out there, at the park. That was exactly what I *didn't* want him to think, especially since he had a direct pipeline with his uncle George. That was a big part of the reason I wanted to get back out there so badly, which is probably also why my mom forced me to promise not to go back out there.

It was sort of ironic, too, because Tommy wasn't much of a physical player, himself. He was more of a 3-point specialist than a post guy. Even *I* had just posted him up, in fact, about a minute before he caught me with the elbow, and like I said, I was a lot smaller than him at the time. So I wouldn't exactly classify the play out there as particularly rough-and-tumble. I just happened to catch an elbow at a bad time.

Anyhow, I was in the first round of cuts from the varsity team, that winter, the same day I won a 3-point shooting contest during the tryouts. Everybody said they were shocked, because they thought I was a lock to make the team. So who knows?

It was nice to get my life back, as it turned out, because there's not much free time left, when you're on the basketball team- especially varsity. I'd have to run track or cross-country, too, if I was on the basketball team, since the coaches want you to do a second sport where you run a lot, but preferably don't get hit. Between that, and weight training, and summer league, it left maybe two or three weeks at the end of summer where you got a break. So we were able to take a nice family vacation early the following summer, for the first time in a long time, without me having to worry about missing a "critical" week of basketball. Plus I was able to intern at an engineering firm, which I never could have done if I was still in basketball. So that was a real nice summer for me, as it turned out, without playing in the league four days a week.

Meanwhile, Merle had gotten a few steps in front of me, and I heard him say, "We should play these two guys in basketball."

I hoped that I misheard him. "What did you say, Merle?" I asked.

"I said, we should play these two guys in basketball."

"That's what I thought you said." I had to almost break into a jog to catch up with him. I was a little panicked because I knew I would be rusty, having not played in almost a year. And my well-intentioned alien friend had *never* touched a basketball, it seemed safe to say. Going two-on-two against Tommy and his friend-- probably a teammate from college or something—would probably end up 21-0 (21 was usually the score limit on the courts at the park), and I wasn't looking forward to being embarrassed like that. Plus, I was terrified that Merle would end up getting hurt. For some reason I had the macabre thought that Merle would get his nose busted, and start bleeding some color other than red.

"Merle, I don't know if this is a good idea," I said. "I haven't played in a long time, and these guys are pretty good players. And pretty rough, too. You've never even touched a basketball in your life."

"I think it'll be fine, Ken. Just give it a chance," Merle said.

"O.K. Whatever." I wasn't feeling very confident about our chances whatsoever, and I had a sinking feeling deep, deep down in my stomach, but I guess I just sort of knew that I'd never be able to talk Merle out of it, anyhow.

By that point, Merle was quickly striding up to the courts, and I was still hustling just to keep up. Tommy and his friend were each lazily dribbling basketballs, and launching up a few practice shots. They each picked up their basketball and turned their heads when they saw Merle approaching. Tommy's friend scowled and said something to Tommy, which we couldn't hear from our distance. But I imagine it was something to the effect of, "now who in the blankety blank are these two doofuses?"

"Hello!" Merle said. "We were wondering if you'd like to play some two on two." We stopped and stood in front of the two. I noticed that Merle was about three or four inches shorter than Tommy, and

Tommy's friend had about two or three inches on me. I was wondering how Merle knew about "two on two".

An uncomfortable silenced followed, as Tommy and his pal assessed the situation with bemused smiles. Tommy's eyes landed on my face, and he recognized me.

"Hey! I remember you! It's the kid with the nose! Long time no see, kid! What brings you back around here after all this time? Finally ready for another beating?" Tommy laughed out loud at his own little joke. "No, seriously, was your nose OK after that?"

"Yeah, I guess so. It healed up."

"That's good," said Tommy, very non-sincerely, I thought.

"We just thought you might be looking to play a game," was all I could think of to say. Tommy's friend whispered to Tommy that he'd rather just practice, and wait for their friends to arrive.

"You can play *us*, in a practice game," said Merle. "*That* will be practice."

Tommy obviously wasn't too interested in what his friend, Harold, thought about it.

Tommy smiled; somewhat sadistically, I thought. "O.K.", Tommy said. "We'll play."

Harold gave Tommy a look that said, "Do we really have to?", while I gave Merle the same sort of look. But we strode out there, anyway, and Tommy went over the ground rules with us, more for Merle's benefit than for mine. Basically, you had to go backcourt with the ball to start the possession, and it had to be checked by the other team after a score. The game was to 21, and you had to win by two points. "Here," Tommy said, and tossed the ball to Merle. "You can in-bounds it to start the game."

Merle took the ball up past the arc, where backcourt was, and started to dribble. I was surprised that he looked perfectly natural, dribbling the ball. "So," asked Merle, "Two points for a regular basket, and three points from behind the arc, right?"

"That's right," said Tommy.

Merle nodded his head. "And I don't have to pass it out first? I can just shoot from right here, if I want?"

"Sure," said Tommy. "We already checked the ball." Those things hadn't been mentioned when Tommy went over the ground rules, and I was surprised to hear Merle ask the questions. A lot of times, in playground rules, a basket is one point, and behind the arc is two points, but the games at Streamside usually went with traditional 2/3 scoring, to discourage players just standing behind the arc and launching shots.

Tommy gave Merle a smile that seemed to say, "Go ahead and shoot, you're going to miss anyhow." Then Tommy stepped across the line, closer to Merle, and put his hands up, in defense of a potential shot. I started cutting cross-court to try and get open for a pass, with Harold at my heels, in hot pursuit. Out of the corner of my eye, I saw Merle just rise up and launch a 3-point attempt, with Tommy's hand in his face, as naturally as if he had been shooting 3-pointers all his life. The shot rose up with a high, graceful arc, and slammed through the center of the hoop. The chain-metal net flipped and jangled, as the ball passed through the basket.

Everybody but Merle just sort of stopped for a moment, surprised. I know that I, personally, was absolutely about as flabbergasted as I could be. In fact, I'm surprised I didn't fall over, right on the spot. Then Merle said, "Three- zip," to announce the score, and the game was on. I didn't even have time to wonder how in the world Merle just did that.

Harold took the ball out at the top of the arc, and as he cut to the right, he passed it back to Tommy as he cut across in the other direction. I was covering Harold, and he cut to the left side of the court after his pass. Tommy flipped an overhead pass across court, leading Harold, who had clearly gotten past me. Harold had an easy layup in the bag, but the ball never got there. Instead, Merle jumped up, incredibly high and amazingly quickly, and snatched the pass out of the air, before it had traveled even six feet. Then he dribbled backcourt, looking a lot like a point guard as he went between his legs to blow past Tommy, while I cut back the other way around Harold.

Merle spun around and hit me with a dead-on pass as I drove the lane for the lay-up. When I went up, though, Harold gave me a subtle push in the lower back, and that little push was enough for me to clang it off the backboard too hard, and off the top of the front of the rim for the miss. As I turned back towards the basket, I saw Merle flying in from the top of the key. He grabbed the ball while it was still up above the front of the rim, after the bounce. While still in the air, he brought the ball over and behind his head, and power-slammed it back through the hoop, just like you might see some guy do in a dunk contest on TV. Merle landed in a crouched position at the baseline behind the basket, after the dunk, facing the court like a tiger, or a martial arts guy, or something. It was pretty wild, really. The entire dunk sequence, even now, when I think about it, is so surreal, how high he got in the air, and how hard he threw the ball down through the hoop, and how he landed and everything. The basketball bounced off the court after the dunk, and Merle snatched the ball back out of the air, while still in his tiger crouch. He looked up at Tommy, who I'm sure was totally shocked by the unexpected start to the game. Merle flipped him the ball. "Five-zip," Merle said.

Well, that move really got their attention, I have to say. Heck, it got *my* attention, also. It quickly became very apparent that Merle, far and away, was the best player out there. Shooting the ball, passing the ball, playing defense, rebounding, Merle did it all out there. Plus, he could jump amazingly high, and, without question, was faster than anybody else out on the court. I had no idea how he was doing it.

We built up a pretty decent early lead, and I even scored a few baskets myself, on an easy layup and a mid-range jumper or two, since Tommy and Harold were so focused on guarding Merle. As the game progressed, it was obvious that Merle was trying to get me more and more involved in the game. It was interesting to watch him operate.

For one thing, if Merle had just taken control of the ball himself, and not passed it to me, we probably would have won the game by 15 points, easily. And when Tommy or Harold made a nice play, or scored or whatever, Merle cheered them on, and wanted to high-five them like he was their own teammate or something. At first they didn't know how to react to that, but after a while, they went along with Merle's routine and accepted the hand slap or fist bump from Merle, laughing at him like he was a crazy man or something. Actually, other than the fact that Merle was passing the ball to me, and guarding Tommy and Harold, you probably might not have even known which team Merle was *on*. That made the game more fun, I think. While Merle was having fun, and at the same time trying to get me more involved in the offense, Tommy and Harold sort of snuck right back into the game. Merle and I had been winning by a comfortable seven points, 19-12, but I missed two easy layups that could have won the game, and Tommy hit a couple of threes to close within one point, 19-18.

Merle took the ball out, and passed to me on the left side. I panicked and dribbled it right off my foot and out-of-bounds.

They took the ball back out, and Tommy passed it back across to Harold, who stepped back behind the arc for the game-winning "three" attempt. Luckily, I moved up on him quickly and got a fingertip on his shot, and it fluttered down, ten feet short of the basket. Merle out-leaped Tommy for the ball and passed it to me, cutting towards the back court.

I took Merle's pass backcourt, and dribbled the ball back out before lobbing it back to Merle at the top of the key. He circled to his left with the ball, and as he did, he gave me a quick head motion, indicating that I should cut down the other side of the lane, which I did. Merle flipped a behind-the-back pass to me as I shot past, and I grabbed it off the bounce. Without taking a dribble, I quickly tossed up a floater. This time Harold's push in the back came a little too late to affect the shot, and the ball hit high off the backboard, caromed high off the front of the rim, and came down straight through the chains, for the game winning shot.

By then, a small crowd had gathered around the court, mostly guys coming out to play with Tommy and Harold, and they hooted up a storm when that winning shot went through the hoop. Nobody would expect Tommy and Harold to lose, two on two, to *anybody*, let alone to us two guys who didn't exactly appear to be serious "ballers".

Merle came over to high five me, and then, somewhat to my surprise, so did Tommy and Harold.

"Nice game, kid," said Tommy. "You hit some nice shots there. Good game. Good, tough game." Then he gave me a fist bump. I even got a fist bump from Harold, who didn't say too much. By now, the four of us had sort of strolled over to the back corner edge of the court, while the other guys came out onto the court and started shooting around. The park was coming to come to life now, as more people poured in, riding bicycles, kicking soccer balls, throwing Frisbees, and walking their dogs. Families were streaming in, and young children were breaking up the early morning peacefulness with shrill shouts and wails.

# CHAPTER
# 20

While Merle, Tommy, Harold and I made small talk, a vendor on a bicycle drove up by the courts, selling ice cold bottles of water for $ 1.50. Merle excused himself from our conversations and went up to the vendor. He handed the vendor a five dollar bill for two waters, and waved off the two dollars in change, telling the vendor to keep it, as a tip. The vendor tried to give Merle a third bottle of water, but Merle thanked him and told the vendor that he didn't need any more water. Merle turned and tossed one of the waters to me. I chugged it in about 15 seconds, I think. Tommy and Harold were drinking their own water, which they had brought with them.

Meanwhile, Merle started to walk back away from the courts, towards the stream.

There was a single tree that was growing on our side of the stream bank in that area. I watched Merle as he walked quickly to the edge of the upper bank, just to the left of the overhanging tree. Merle set his unopened water bottle down in the grass, next to the path, and he began to hop deftly down the bank. He stopped along the way to grab a long branch that was lying amongst the tangled heaps of weeds, vines and assorted trash, and then he continued on his way, down to the very edge of the stream. Mystified, I followed over a little closer, to get a better view of the situation.

A child's ball was in the flow of the stream, and it was rapidly approaching Merle's location. Even though he seemed perilously close to falling into the brown water, Merle reached the branch outward, and, with the gnarled and curved end, he corralled the colorful sphere as it floated by. It was one of those plastic, inflated toy balls that you get for three dollars in some grocery store bin, about eight inches in diameter, cobalt blue, and festooned with several emerald green cartoon characters with brown pants and cottony white hair. Merle was just in time to grab the ball with the stick, and he pulled it back to where he stood on the bank. I was surprised by how he knew the ball was floating along back there, and even more surprised at how quickly he moved to make the save. I was confused as to why he went out of his way like that, just to grab a little child's ball. I was thinking that maybe Merle was going to collect it, to bring back with him on the ship, as something like an artifact, or a souvenir, maybe. I was starting to wonder how I was going to try and explain this odd behavior to Tommy. This was really the first goofy behavior which I had witnessed from Merle.

Meanwhile, Merle hopped back up the river bank, pausing along the way to drop the branch back into its original position in the tangles. At the same time, I started to see what had actually happened. The commotion of a wailing kid was coming closer, and I turned to my left to see a young girl, no older than six or seven years old, with tears streaming down her face. She was running along the path by the stream, looking down into the water, apparently searching frantically for her ball. The instant she spotted Merle with the ball, she stopped in her tracks, while still wailing and crying. Her thick curly locks of hair bounced up and down with each heart-rending sob. Her eyes were locked on Merle, as he rose up from the riverbank, with the ball in his hand.

The little girl's eyes followed Merle as he strode over and picked up his water bottle. He opened the bottle and set the cap back down in the grass. Then he cradled the ball in his left arm and poured some water onto the ball. He managed to swoosh the water over the surface of the ball with his right hand, while he sort of pinned the ball against his chest with his left forearm. He shook the water off the ball and then poured out a second round of water and swooshed it across the ball again. He shook the water off and finally emptied the rest of the water onto the ball, swooshing and shaking off the water a third time.

He reached down to pick up the bottle cap, and he quickly screwed it back onto the empty bottle. Then he turned to his left and pitched the bottle into the nearest recycling bin, which was about ten yards away. The bottle went in, "nothing but net", of course. Then, holding the ball with both hands, he carefully rolled it back and forth across the front of his shirt, for a final drying.

Merle looked up, and he met eyes with the little girl, who was still standing on the path, maybe twenty yards upstream. Another forty yards back, the girl's mother was closing in, chasing down the path with quick, scurrying steps. Every few steps, the mom would pause and sharply call out to her wandering daughter, in breathless, staccato bursts of some foreign language that I couldn't place. Farther back still, the little girl's brother was getting vigorously scolded by their rather angry father. Apparently the brother had kicked his sister's ball into the stream.

Merle held the ball up in his right hand, showing it to the little girl. She instantly stopped sobbing and stared at it with wide eyes. The sun had now moved up higher into the sky, above the tree line, and it brightly illuminated the blue globe that had been lost, and now found. Merle, in perfect bowling form, rolled the ball slowly down the path, back towards the little girl.

She stood there, transfixed, and watched the rolling of the approaching ball. Finally, it curved back across the path and came to rest, right at her feet. She reached down and picked up the ball. Cradling it tightly in both arms, she peered intently at Merle, for a good few moments. Then, suddenly, she broke out in an immense grin of acknowledgement and squeaked out a quick "T'ank you!" in Merle's direction, before she turned on her heels and ran back down the path, jabbering excitedly to her mother in that unknown language. After the girl reached her mom, and showed her the ball, the mother looked across at Merle. She held up her hand towards him and smiled, as an obviously very sincere and respectful "thank you". With her arm around her daughter's shoulder, and with both talking excitedly, they turned and walked back towards their picnic spot. I suppose they might have moved their picnic a little farther from the stream, after that turn of events.

"You're welcome!" Merle called after them. He was smiling broadly. Then he turned and strolled back over towards the courts, to where Tommy and Harold were. I walked over there, also, relieved that at least there was an explanation for why Merle went for the ball.

"Nice job, man!" Tommy Marnel said. He thumped on Merle's back repeatedly to show his enthusiasm for what he had just witnessed. "Nice job! I can't believe you saw that happening! That was beautiful!"

"Why did you pour the water on it?" Harold asked Merle.

"That water in the stream is filthy," said Merle. "There's a lot of garbage in there, and some of the factories upstream seem to think the stream is their own personal sewer.

So I just wanted to clean off the ball a bit."

"Oh," said Harold. He seemed very satisfied, and even impressed, by that answer. "Right."

"There is a lot of garbage in there," I said. "Shopping carts, old tires, plastic bags, all kinds of stuff."

"It's pretty bad," Tommy agreed.

"Something ought to be done about it," Merle offered.

Harold laughed. "Right! They can't even manage to empty the garbage cans out here, half of the time, and they're going to take the carts out of the river!"

"But why wait for the Park District to do it?" Merle asked. "Why not get something organized on your own?"

"I don't think they'd let us," I said.

"Sure they would!" said Merle. "You'd just have to get some sponsors."

"Well, if you ever do it," said Tommy, "let me know. I'll get a whole bunch of people out here to help."

"O.K." I said, laughing. "Fair enough." I looked over at Merle. "We really need to get going," I said.

"Well," said Tommy, "thanks for a good ass-whuppin' out there." He was looking at Merle as he said that. "Where did you play b-ball?" he asked Merle.

"Nowhere, really," said Merle. "Just here and there, basically."

Both Tommy and Harold laughed at that one. "Right!" said Tommy. "'Just here and there!' Right!"

Merle laughed also, and we all shook hands. Merle and I walked away, while Tommy and Harold went back to the court, shaking their heads.

# CHAPTER
# 21

"Wow," I said to Merle, as we walked away. "That was unbelievable." I was referring to both the basketball game, and Merle's rescue of the child's ball.

"Cute kid, isn't she?" said Merle.

"Very cute kid," I said. "What language was that?"

"I don't know yet," said Merle.

Merle told me what the language was, later that day. But I didn't recognize the name of the language, when he said it, and I don't remember it, now, or even the country of origin. But it doesn't matter, anyway.

"How in the heck were you able to play basketball like that, Merle? You've never even touched a ball before!"

"What do you mean, never touched a ball? We have a court on the ship. I play just about every day."

"What the heck are you talking about? You have a court on the ship?"

"After I first started coming down to the planet, about five years ago, I became interested in basketball. It's actually fairly similar to a sport we play on Akeethera, in some respects. So I requested that they put a basketball court on the base ship, so we could play in our spare time. Since then, they've also put in volleyball courts, batting cages, and more. Many of us enjoy playing some earth sports, when we have some free time up there. I've been trying to get them to put in a hockey rink and a couple of bowling lanes, also, but I don't think that's going to happen. I guess they have to draw the line somewhere. I can especially understand not getting a hockey rink, I suppose."

I was quite taken aback by this information, although I had already witnessed the immensity of the ship for myself. Before I could say anything more about it, though, he continued.

"By the way, Ken, you played great out there. Especially considering you hadn't played in quite some time. What a great finish, deflecting that shot and then making the floater at the end." Merle allowed a smile to spread across his lips. "I can't believe how satisfying that game was, Ken. How simply satisfying it can be to win a game, under certain circumstances… How very interesting and amazing! I'm sure I haven't felt that way since I was a very young boy."

Before I could ask what Merle meant by that, he suddenly stopped and bent down in the grass alongside the side of the path, looking at an ant colony, apparently.

"*Formica subsericea*, I believe. So beautiful," Merle said, on his hands and knees, peering closely at the ants. "Quite remarkable."

"Ants are pretty amazing, I guess," I said.

"Yes they are," said Merle. He was peering closely at a particular scurrying ant. "Or is that *formica pallidefulva*? It can be so difficult to tell the different species apart!" Merle actually sounded a little exasperated, and he stood back up.

At that moment, Merle saw something out in the distance, and he pointed across the park. "Look! We're in luck! Ha ha!" Merle was pointing across the lawn at a man approaching on a bicycle. He was on the path, about 200 meters away from us at that point, and headed our way. "This is tremendous!" Merle exclaimed loudly.

To my surprise, I recognized the cyclist immediately. He was a well-known neighborhood character. My friends and I had seen this shabby-looking guy, with his scruffy white beard, riding his bicycle all over town, countless times; ever since we were little kids. He rode in all kinds of weather- rain or shine and sometimes even snow.

Everybody always said that he was homeless, and we never saw him driving a car—just his bicycle-- so we called him "Homeless Bicycle Guy".

I remember that Seth Millingham once spent an entire summer collecting crabapples, with the stems pulled off, so that he always had a few in his pocket, in case Homeless Bicycle Guy happened to roll by on his bicycle. Seth would fling a few crabapples at the spokes of Homeless Bicycle Guy's wheels as he passed by, hoping to get a good "ping" off of the spokes. If he managed to pull it off, we'd all howl with laughter, and talk about it for the next week and a half. Homeless Bicycle Guy pretty much ignored us, once he figured out that we just wanted to ping crabapples off his spokes.

Homeless Bicycle Guy even became a sort of meme with our group, at some point. One of our favorite little jokes was that if something less than desirable happened to somebody in our group, he would say something to the effect of, "Well, that really sucks, but at least I'm not Homeless Bicycle Guy!" Then we'd all roar with laughter. People used to say that Homeless Bicycle Guy lived under a bridge like a troll, or that he took baths in the river, and all kinds of disgusting stuff like that. I always assumed it was mostly true, since everybody always said it.

I was very confused as to how Merle knew Homeless Bicycle Guy, and why.

"Do you know that guy, Merle?"

"Why yes, I do. Quite well, actually."

For a moment, I had the stunning thought that Homeless Bicycle Guy wasn't at all who we thought he was. "Is he an alien, too, Merle?"

"Certainly not. He grew up right here in this area."

"Isn't he homeless?"

"Most certainly not."

"Are you sure?"

"Of course I am sure. I've been inside his house. And what would if matter if he was, in fact, homeless?"

I ignored Merle's question. I was a little stunned to hear that Homeless Bicycle Guy wasn't homeless, after all. And in some strange way, I was a little disappointed to hear that he wasn't an alien, either. "You've been to his house?"

"Yes, I've been to his house," Merle said. "He has a very nice house. And a very interesting insect collection."

"An insect collection?" I wasn't sure if I had heard Merle clearly.

"That's right. Charles is a professor of entomology. Charles A. Jonmur, Professor of Entomology. Although he is retired, now."

"Oh." I said. I wasn't exactly sure what an entomologist was. And I still couldn't believe that Merle knew so much about this "Charles Jonmur".

"Entomology is the scientific study of insects," Merle said helpfully. "Unlike etymology- that is the study of the origins of words or linguistic forms."

"Oh, that's right. I knew that." At least, it sounded vaguely familiar.

"In fact, Charles has taught at some of the top universities in the country, and he's worked for a couple of the best natural history museums, also."

"He has?"

"Oh, yes. He is extremely knowledgeable, extremely so."

Well, it seemed like new information was pouring in, left and right, about old Homeless Bicycle Guy. I was taken aback by the whole thing, no doubt, and I had a series of additional questions to ask. But before I could ask another question, Homeless Bicycle Guy— or Professor Charles Jonmur, I guess—was upon us.

"Charles, I'm so glad to see you!" Merle was obviously happy to see his friend, as he rolled up to us and nimbly hopped off of his bicycle. "I need your expert opinion here!"

As Professor Jonmur set the kickstand of his bicycle, I got a good close look at his attire, which I had always thought of as hideously shabby. After all, this was a man alleged to live under a bridge, and bathe in the river. But now, from up close, and perhaps with a more sophisticated eye for clothing compared to when I was a pre-teen, I could see that there really wasn't anything shabby about his clothing. In fact, his clothing was more up-scale than my own clothing, I was quite sure. It certainly appeared to be impeccable, especially for a man reputed to live under a bridge. Even the "scruffy" white beard appeared to be well maintained and neatly groomed, for the most part. Bubbles were bursting for me, left and right, and I was already quite doubtful about the entire "living under the bridge" story.

Merle extended a handshake to his friend, and introduced me to Charles. "Ken, I'd like you to meet my friend, Professor Charles M. Jonmur. Charles, I'd like you to meet my friend, Kenneth Sylvanewski." Charles and I shook hands and said "hello." I experienced an initial horror, as I was afraid the professor would recognize me as a member of the old crabapple gang. Fortunately, he did not.

Merle didn't waste any more time on the introductions. "Charles, I have a question for you." Merle gestured down towards the ant colony. "Am I looking at *formica subsericea*, or am I looking at *formica pallidefulva*?"

Charles strode over, bent his knees and crouched low to the ground, peering at the ants. "Ah!" He turned around and looked at Merle. "Neither! It's actually *formica incerta!*"

"*Formica incerta*? Oh, of course! I should have probably known that, from the location alone." Merle turned and looked at me. "It's a colony of the 'Uncertain Field Ant'. They usually inhabit more natural settings, like this rough area near the wooded edge. They can forage in an area that doesn't get herbicides applied regularly."

"That's right," said Charles. "It's a bit more hairy, also, than the 'Slender Field Ant', or even the 'Silky Field Ant'." Charles was a little unusual, as an entomologist, in that he wasn't aghast at the idea of using common terms for insects, if he was in "mixed company". I'm sure he had already picked up that I wasn't anywhere close to being on Merle's level, as far as entomology.

Merle was absolutely beaming in satisfaction. "Well," he said. "I'm so glad you were here to clear this up, Charles! I was absolutely lucky on this one! *Formica incerta*, I should have known! Wonderful!" He bent down again, to peer more closely at the ants.

"If you went across the street from the park, and looked at some of the front lawns of some of the houses there, you'd probably have a better shot at seeing a colony of *formica pallidefulva* or *formica subsericea*," Charles said helpfully.

Merle laughed at that. "I could see myself getting arrested for that! Man is arrested for looking on peoples lawns, for ants!" We all laughed at that image.

"It might be better if we just stay out here, in the park," said Charles.

"True, true," Merle chuckled. "Speaking of the park, did you know, Ken, that this park is named after Charles' father, Maximilian?"

"Really?" At this point, I wouldn't have been surprised if he told me that Homeless Bicycle Guy, a.k.a. Charles Johnmur, was the mayor of Rundle Heights.

"That's right. The actual, full name of the park is 'Maximilian R. Jonmur Streamside Park'."

"Why isn't it listed on the sign like that?" I asked, looking at Charles.

"Well, I'm afraid it's a bit of a long story," Charles said. "Basically, my father was the Village Manager of Rundle Heights, back 50 years ago or so. This park was his idea. You see, he knew that this vacant parcel was valuable habitat for fish and other wildlife, with space for sporting fields, also, and that it connected with the forest preserve that stretches all the way to Baxter Lake."

"Baxter Lake! I know that lake! I didn't know that the forest goes all the way to Baxter Lake."

"Well, yes, it does. In fact, the stream originates from Baxter Lake, which is a spring- fed lake. Of course, after they redesigned the park, a lot of wastewater is fed into it, also, before it reaches this area. They added the channel from the sanitary canal, too, so now it's mostly non-spring fed water, by the time it passes through Streamside Park."

"Baxter Lake is spring-fed? I did not know that!"

Charles extended his arm out and panned around the park. "Do you see how this land is basically a giant rhombus?"

"Oh, yes. I see it." I had never really looked at the shape, but it was, as Charles said, essentially a giant rhombus. All four sides were pretty equal in length; it was basically a large square that was slanted off to the side, into a rhombus.

"Back before this was a park," Charles explained, "this land was bisected. Basically, the southern half was a raised, level plateau, perfect for playing fields, and that sort of thing. The northern half was a sunken down area, where it would be sort of swampy in the spring. The northern part tended to dry out in the summer heat, but after a big storm, you'd see it all spring right back to life. The stream ran along the edges, but it was very shallow back then, and at parts it sort of fanned out into little streamlets that wandered off into the flood plain area during the wet times. Most of the time, you could step all the way across, without getting wet.

"My father thought that it would be a perfect combination facility. The flood plain would serve as a sort of nature reserve on the northern side, while the raised southern portion would be perfect for playing fields, and that sort of thing. My father envisioned a boardwalk circling through the northern side, where the cattails used to be, so people could go there even during the flooded times. On the raised southern section, his idea was to build four baseball fields, with basketball courts, tennis courts, fields for football-- which might be soccer fields, these days-- a playground, a concession stand, and maybe some other things like horseshoe pits, tetherball, or whatever else may have been popular back then. He wanted to put parking along the south portion of the western border, and another parking area along the southern perimeter."

"Isn't that sort of how it is now?" I asked, rather stupidly, as it turned out.

"Not at all! This isn't anything at all like what my father envisioned." Charles frowned. "The neighbors along Cypress Avenue on the southern border didn't want a parking lot on their side, so instead the parking lot got built where it is now, on the east side, along Basin Street. At the time, there were no houses on Basin, so there were no neighbors to complain about it."

"But everybody just parks on Cypress Avenue now, to get to the park." In fact, that's where we had parked. "Wouldn't the people rather have cars parked in a lot, instead of on their street?"

Charles smiled ruefully at that. "Maybe, who knows? Now the neighbors along Cypress complain about all the cars on the street, and the people on Basin complain about the parking lot, mainly because half of it is in the floodplain, and it's very difficult to maintain because of that."

"Is that why the parking lot is always all broken up?"

"Yes. The northern end, at least, is always breaking apart. Same with the old concession stand, that should be torn down. That's because the men who constructed the park didn't want to listen to my father." Charles stopped speaking and looked around the park for a few moments, and I uncomfortably realized that he was getting a little emotional, and was trying to regain his composure. He continued, though.

"They decided to level out the entire southern section, removing all that rich topsoil, to sell for peanuts, no doubt, and they dredged out and re-routed that beautiful little stream into the sewer of a drainage ditch that it now is, along the western and northern edges of the park. All the little streamlets, where I used to catch frogs and minnows as a child, were obliterated. They built three baseball fields out here, on the north end, by the stream. That's why the fields always flood in the springtime. They thought they could just put up a berm, and keep the water out. But water had been flowing into that area for hundreds, if not thousands, of years. It's not so easy to keep that water out."

"Foul balls always end up going into the stream, too," I said. "We used to call it "Swampside Park, when we played here as kids. Always getting rained out, even if it hadn't rained in a week and a half."

"That's why my father wanted to put all the playing fields up on the plateau area, on the south side of the park. No flooding problems up there. In fact, the drainage on the south end used to be excellent. The maintenance costs would have been miniscule, compared with how it is now."

There was another uncomfortable silence, and I thought I should say something. "It sure sounds like they made some bad decisions."

"Yes. That's why my father insisted on not having his name on the sign, since the day it opened over 45 years ago. He always deeply regretted what had become of the park. That's why, to this day, so few people know what the full name of the park actually is."

"So the stream used to dry up in the summer?"

"Well, the stream would sometimes dry up almost completely before it got down to this area, if we had a long hot patch of weather. That's when you could walk around down there and see toads, and all kinds of other critters and insects, without having to worry so much about the water.

"The park planning commission didn't want a stream that dried up in hot weather, though. They said that they wanted to use the stream as a flood preventative measure, since it flows all the way down to the Blackhawk River. That's why they linked it up with the sanitary canal. If you follow the water, it eventually feeds all the way down to the Mississippi, and out into the Gulf of Mexico, ultimately.

"It would have been a better flood deterrent if they had left the stream the way it was, when the land absorbed a lot of the excess water during wet seasons. Now, it's just a channel that carries the water downstream, where it becomes somebody *else's* floodwater. And the quality of fish habitat was almost completely degraded when they 'channelized' the stream, as well. You'd be hard pressed to find ten fish in there, now. 50 years ago, this area was literally teeming with fish, frogs and toads."

"There were frogs and toads out here?"

"They were very plentiful, very plentiful. Also, there used to be salamanders, but those have been gone for at least 30 years, probably.

"There were lots of turtles, too. You had to watch out for the snapping turtles, though!"

"Wow. I would have loved that, as a kid-- except for the snapping turtles."

We all stood there a moment, looking about, and visualizing the way it used to be, just 50 years prior.

"It's a shame, certainly," said Merle. "Disgraceful. They wasted an inordinate amount of money "improving" things that should have been basically left alone, and now money is again needed to repair the damage. Actually, Ken and I were just talking about the stream; how polluted it is, and how it would be great to organize a cleanup."

Charles instantly looked more hopeful, at the mention a cleanup of the stream. "A stream cleanup? That sounds like a fantastic idea! And also well-needed! Then, maybe we can rebuild the southern section, move the fields and courts back there, and re-engineer the stream, back to how it was originally!"

"Now we're really getting carried away!" I said. "It would be a miracle just to see the shopping carts dragged out of that stream, let alone re-shape the whole thing!"

"I don't think it would quite take a miracle," Merle said.

That conversation turned out to be the seed for a great friendship that I eventually developed with Professor Jonmur. I never told him about being in the Crabapple Gang, though, until I was halfway through writing this book. We had a good long laugh about that! The Professor even helped me verify the scientific names for some of the insects and animals that Merle had mentioned; otherwise I never could have remembered what he called that cicada, or cricket, or whatever.

To this day, Charles is very non-committal, and non-judgmental, about where Merle came from, exactly. I think it is enough, for the Professor, that Merle was truly an entomologist and a conservationist at heart, whether he carried a degree, or not, or whether he was truly an alien, or not. Although I suspect that, deep down, Charles knows who Merle really was. He just finds it hard to admit, publicly, as a scientist.

# CHAPTER 22

We parted ways with Charles, and Merle and I continued along. A crow—*corvus brachyrhynchos,* according to Merle-- was high up in a tree, chattering about a man who was sitting on a park bench down below, tossing peanuts in the shell to a squirrel. The crow flew over to the top of another tree, which was the highest tree in the area, and began cawing loudly in all directions.

"See that?" Merle asked, pointing to the crow. "Calling to his buddies."

I had no idea what Merle was talking about.

Merle bent down to check out another anthill. "I guess that's why I'm so interested in the ants. They also cooperate very effectively, in a group. People, on the other hand? Sometimes a little hit or miss." Merle chuckled at that, and he stood back up, craning his neck again to watch the crow. Sure enough, in a few moments, another crow showed up, followed by two more, a few moments after that. Before long, each of them had acquired a peanut or two from the man on the bench, who seemed to know his feathered friends quite well from previous feeding sessions.

"They have trained him well," Merle observed. "And all five seem to be enjoying it. Six, if you count the fox squirrel." Merle had already thrown down the scientific term for fox squirrels, earlier, so he now referred to them simply by their common name. Although he still had to break down specifically which variety of squirrel it was.

Heading back to the car, it was hard to believe that only about two hours had passed since we arrived at the park. I was amazed at all that had transpired out there, in such a short amount of time.

After we collected our last item of garbage, an old broken and discarded Styrofoam cooler that had been tossed into a bush at the edge of the parking lot, we got back into the vehicle. Merle touched the front screen, and music came on. It was "Green River", by Creedence Clearwater Revival.

"Man, I just love this song," said Merle. "You know who it reminds me of?"

"No, who?"

"William Wordsworth."

"Who's William Wordsworth?" That was one of those moments, when you say something that doesn't seem dumb at all, when you say it, but then afterwards, you still feel sort of dumb about it. But it's too late, then.

"Who is William Wordsworth? He was probably the greatest poet of the Romantic Era of English Literature! He was one of the all-time great revolutionaries of Western Literature! This song couldn't even *exist* without Wordsworth!"

I sure wish I could somehow have a picture of the expression on my face at that moment. I had no idea what Merle was talking about, basically. "I don't think I've ever heard of him," I said.

Merle, blissfully unaware of the irony, looked at me like I had just floated down from outer space. "That is very surprising to me," he said. Then Merle turned up the volume a bit, on the song. "Let's go see what's up around the bend, shall we?" Merle asked.

"Sure."

Merle pulled out of the spot and spun a U-turn on Cypress. We headed back east and turned left on Basin, heading north, back towards the basketball court area. We pulled into the parking lot, and Merle stopped the car behind the old concession stand. At this point, the stand blocked our view of the homes on Basin, and a stand of bushes blocked our view of the courts. "Here we go," Merle said.

All of the sudden, the car-- or, the ship-- lifted off the ground, and we floated up, above the tree top level. "Merle, everybody can see us!" I shouted out.

"No, they can't. We're totally cloaked."

"We are?"

"Absolutely." To prove his point, Merle brought the ship back down, below tree top level, and floated across the park to the basketball court area. Sure enough, nobody seemed to notice the craft in the air as it floated over the area. We settled right over the court, where the last game had just ended.

Tommy and Harold's team had won, and some of the guys on the other team were giving them a hard time about being "only 1 and 1" on the day. "I can't believe those guys beat you!" I heard one guy say, quite clearly. Merle turned to me and smiled, and then we lifted up higher, above tree top level again, and headed over towards the stream.

I looked down into the murky water, and suddenly we swooped down, very low, about eight feet above the water. There were several shopping carts down there, plus all the other assorted papers and plastic that we had noted earlier. It was actually much worse than I had even imagined. "Wow," I said. "Sick."

"That's not the worst part," said Merle. "Let's head back upstream a bit." We flew along, following the course of the river, until we got to the section that was bounded by factories on the north side of the stream, just after where the water from the sanitary canal entered the stream. Merle scowled. "Some of these factories are good, decent factories," he said. "But some of them like to dump their foul messes into the river."

I was sort of shocked to hear that, and Merle could see it on my face.

He seemed surprised at how naïve I was about the industrial wastes seeping into the stream. "That truly surprises you?"

After thinking a moment, I realized that it probably was par for the course. "Well, I guess I shouldn't be too surprised," I said.

"No, you really shouldn't, unfortunately," said Merle. He pointed to the right, and there, in the back yard of one of the factories, maybe two or three meters from the river bank, were about three dozen large, rusted out metal barrels. "Most of those barrels right there are leaching caustic materials into the river. Every time it rains, more materials flow down into the river. That's been going on for at least five years now."

"Crap," I said.

"Crap indeed," Merle said. We floated towards the adjacent factory, and Merle pointed out a long, narrow pipe that ran from the building, to nearly the edge of the stream. "That pipe carries water away from the factory-- waste water, contaminated with heavy metals, and other toxic chemicals. It leaches out of the stream bank, and goes right into the stream. Notice there's no grass growing around the outlet of the pipe. There are several of those pipes along the riverbank."

"Crap. Doesn't anybody check up on these places?"

Merle just looked at me, blankly. "What do you think?" he asked. "Of course there are agencies, but they are rather chronically overwhelmed."

We followed the stream, all the way to Baxter Lake. The portion that ran through the forest preserve was still a nice little stream, mostly in its natural form, still.

"You *did* want to see the lake, didn't you?" Merle asked. "Yes, thanks. It's very nice from up here," I said.

"Yes, it is very nice." We looked down at the lake. Fishermen stood here and there at the water's edge, minding their lines. Canoeists were paddling along, ducks bobbed up and down on the waves, and gulls circled overhead. I wouldn't have minded circling around there a little longer, but Merle had other ideas. "I know another nice place, too," he said. "Let's go check it out." And with that, we rapidly began to ascend, high into the air. I reflexively held onto the arms of my chair, without needing to. Although we were moving fast, I couldn't really feel any movement like you might expect. It was as if we were stationary, and the scenery was streaming past our stationary post. I relaxed my grip and peered out the window. We were already high enough so that the cars below were as small as ants, as they say.

As we continued to rise even higher, I noticed that the appearance of the ship's dashboard changed. Two of the screens that were there, previously, blinked out of existence, and several additional screens appeared in the same general locations. One of the screens in the middle even assumed a sort of shell-like, three-dimensional shape, like a hologram. Dozens of points of light were suspended within the upper regions of the shell, above a scaled-down view of what I took to be the Midwestern portion of the United States. I could see Lake Michigan at the top, or north, of the shell. I rose up a bit in my chair to look down out of the front windshield, and I was surprised that we were now high enough in the air to see what appeared to be, in fact, much of the Midwest.

"Oh my gosh! How high up are we, Merle?"

"Not too high, yet. Still within the lower atmosphere."

"Still within the lower- where are we going, Merle?" I was suddenly gripped with panic. I realized that we could easily continue to sail upwards, out of the atmosphere entirely, and on to who knows where. I had that feeling, again, of being taken for the ultimate ride with a stranger.

"Don't worry, Ken. We're not going to leave the planet. We're just taking a little sightseeing tour, that's all."

I looked down again. "Is that the Mississippi down there?"

"Yes, it is. There goes some of your water, from the stream that leads out of Baxter Lake. All the way to the Gulf of Mexico."

Down to the bottom, or the south, of the screen, a large body of water came into view. "Oh, wow, is that the ocean down there?"

"Yes, it is."

"Wow, wow, wow. This is way higher up than airplanes fly, isn't it?"

"Oh, yes. Much higher."

"Wow! Where are we headed?" It looked like we were turning to the east, now.

"You'll see."

I looked at the shell-like holographic display again, and I saw that there were many more points of light, floating up above a large area of ocean, with the southeastern coast of the United States, as well as Cuba and Puerto Rico and other islands of the Mediterranean, in view. A vast area of the North Atlantic Ocean began to occupy an ever-greater portion of the screen, and up ahead, another coastline was coming into view. "What are all those floating lights up above the ocean, Merle?"

"Those are ships."

"Ships on the ocean? Why are the lights in the air, then?"

Merle laughed at that. "No, I'm sorry, not ships on the ocean. Ships in the air!

Spaceships, I suppose you would call them."

"Spaceships! You've got to be kidding me! How many are there?"

Merle moved his finger towards the dashboard of the craft, and another, smaller, flat screen appeared. He jabbed at it several times, and some very hieroglyphic-looking figures appeared. "Right now, about 75 or 80, on the screen."

"75 or 80!" For a moment, I truly, truly began to panic. "Merle! Is this an invasion? Are they all cloaked?"

At that, Merle turned and looked at me. Then he burst out in laughter. "No, it's not an invasion, Ken. And yes, they are all cloaked. That's how it always is."

"Why are they here, then?"

Merle gave me one of those "isn't it absolutely obvious?" looks. "Well, to learn about the Earth, Ken. You're living in very interesting times on this planet, and people want to see how things unfold. Also, by observing this important part of Earth's history, we all learn about the past on our own planets, and so we all learn a little about ourselves. At some point in the distant past, we were all much like Earth is today- fighting and killing each other, struggling for food, searching for shelter, and damaging our planet, while beginning to explore the cosmos."

"Where do they all come from? Are they all from Akeethera?"

Merle laughed. "No, no. Our big triangle is the only ship out there from Akeethera, other than a couple of dozen scout ships based on the triangle, and a few on-planet ships, like this one. All these other ships come from all over the galaxy, and many are from different dimensions of the galaxy. A handful of them are even from outside of the galaxy, entirely."

I let the comment about "different dimensions of the galaxy" slide. "So why don't they just come down and say hello? Why do they hide?"

Merle gave me another of those "isn't it obvious?" looks. "That would be the worst thing we could do, Ken. The people of Earth are very xenophobic and frightened, and they are not ready for that. Too many would panic and spew fear, unfortunately.

"On the other hand, there have been discreet contacts and overtures made, many times, on a limited basis. Our mission, for example, involves discreet contact. And our mission may possibly lead, eventually, to a more significant disclosure, depending on the results."

I thought back to one of Merle's previous comments. "Do you mean to tell me that there are always 75 or 80 freaking spaceships floating around up here?"

"Well, just in this one area, sure. I mean, usually there are nearly a thousand, around the world, at any given time. That's a little misleading, though. A lot of the ships are small-- scout craft, not much larger than this ship that we're on. There's probably only about 50 or 60 large base ships, like our triangle, worldwide. Most of these smaller ships you see are based out of the larger ships." Merle peered more closely at the shell-like screen. "A few of these blips are larger ships, though."

"50 or 60 of those big triangles?"

"Well, not all are triangular. But in general, yes! Sometimes more than 50 or 60, even." Merle was clearly enjoying my astonishment. "Then there are the very large ships- what you might call the mother ships. Those are still much higher up, out of the limits of the screen."

"The mother ships are up higher than this? Why?"

"It takes more effort to cloak the mother ships, since they are so large, and there's no need for them to be so close, anyway. It's easier for them to hide further away. Some are on the other side of the moon, and others are scattered in different areas throughout the solar system. Most of them are probably exploring around, out by Jupiter or wherever. They move back closer in, if they need to."

"Throughout the solar system!" Not surprisingly, I was quite astonished at this string of revelations. I looked at the shell-like screen again, since I had already learned that it wasn't necessary to physically look out the window. We were now over a large landmass—apparently a desert, for the most part. The ocean was now farther off behind us, and patches of water were off to our left, and in front of us. Looking out the window, I could see that the sun was now off to our west. "Where the heck are we now, Merle?"

Merle peeked at the shell-like holographic display. "North coast of Africa. Probably crossing over Egypt, and about to cross over the Arabian Peninsula." He looked over and saw the astonishment on my face. "Pretty wild, huh?"

"I'll say. I can't believe this. How fast are we going?"

"We're not really traveling very fast at all- nowhere near what one might consider to be relativistic velocities. Still, we're fast approaching our destination."

Looking at the screen again, I could see that night was beginning to settle across the land below.

"It's nighttime out here."

"We're on the other side of the world, now."

"I thought you were supposed to stay in the U.S."

"This is O.K. We're just going to visit for a while. There is a lot more to the world to be enjoyed than just the U.S., my friend."

With that, I peeked out the window again, and saw that although our ship was still up in the sunlight, the land below was more deeply shrouded in shadow. The details of the land below were still clearly visible on the shell-screen, though. There were more desert-like terrain below us, with patches of water around us, including the large expanse of ocean to our South, which was on the right side of the screen.

Merle pointed towards the ocean on the screen. "See the Arabian Sea down there?"

"Is that what that is?"

"Yes. It's basically the northern portion of the Indian Ocean. Water from our little stream ends up all the way out here, also. Some of that water will come down as snow, on top of the mountains to the east. Of course, there is plenty of pollution being produced in this area, as well, just as it is being produced in every industrialized center of the world. That pollution will also circulate through the oceans and atmosphere, and some will end up in your own backyard, and back into the water and the air at Streamside Park. That's why there truly are no completely unspoiled areas left on this planet."

"Damn," I said, both in wonder at the scene, and at the thought of the relentless cycle of pollution, as Merle described it.

I looked east to see mountain ranges, now clearly in view, and I realized that we were beginning to descend. By now, our ship was also in the shadow of the earth, as night fell completely on the landscape below. The patches of desert-like terrain were quickly rolling away to our west, and directly below us, now, were mountains. A larger, vast chain of mountains was rapidly approaching in front of us, from the east, as we continued to descend. I looked out the window again and was startled to see an unbelievable number of stars erupting out of the darkening sky above. "Wow," was again all I could say.

"That's nothing. Wait until we land."

"We're going to land?"

"Sure! That's why we came out here."

I looked at the screen again. We were still descending, and by now the screen was filled by a massive chain of mountains, stretching out seemingly into infinity, below. "Oh my gosh! What mountains are these? They're huge!"

"That, my friend, is the Himalayan Range."

"Wow." I don't know if I could adequately express the visual impact of the Himalayas, seen from this perspective. Awe-inspiring, majestic, glorious, magnificent- none of these words come very close to describing the actual scenery. I found myself, quite literally, with my mouth wide open as I scanned the monumentally vast, snow-capped range.

"Amazing, right?" said Merle.

"Amazing is right. Oh, my, gosh. Amazing, amazing, amazing."

"I know. This is my favorite spot to go and think. This is actually one of the most awe-inspiring places I have ever been, anywhere in the galaxy."

"What country is this below?"

"I don't really want to say exactly where we are going to land. But I'll go anywhere from Northern India, to Nepal, to Bhutan, to Western China. It's all beautiful. If I'm looking for animals, I might head more towards the Tibetan Plateau."

I gripped the armrests more tightly, as the mountains below were zooming towards us as we came in for a landing.

"Don't worry," said Merle. "I've done this almost a hundred times, now."

And with that, I realized that we were already on the ground. Or, more specifically, we were on the snow, on top of a small plateau, near the top of one of the higher mountaintops in that area of the range. I looked out the window, and by the surprisingly bright starlight I could see, rather clearly, innumerable gigantic mountains reaching for the sky, all around us. I reflexively whistled in amazement. Then, Merle touched his finger to the screen, and the entire ship faded away into transparence, leaving just Merle and me, in our chairs and still perfectly warm, sitting on a mountaintop somewhere in the Himalayas, in the middle of the snow. The stars above us were so overwhelmingly numerous, that I audibly gasped. "Oh my gosh. Oh my gosh. Oh. My. Gosh. Merle, it's so beautiful up here!"

"Yes, I know."

"Where is the ship?"

"We're still in the ship. I just like to get rid of the clutter, so we can see better."

"Awesome, Merle."

"Yes."

# CHAPTER
# 23

Merle gave me some time to soak it all in. He went back to the dashboard, temporarily made visible, and tapped it. The sounds of the high mountain wind came through, very loudly and clearly, as it howled ferociously over the rugged and unyielding landscape. At that volume, we'd have had to shout, just to hear each other. Merle continued to tap the dashboard, and the sound reduced down to a more manageable, almost peaceful level. Then the dashboard disappeared again. It truly was like sitting on a couple of chairs on top of the mountain, with no other indication that there even was a ship in the vicinity, except maybe for the fact that we were still warm, and the sound of the wind was subdued. I turned in my chair, which turned right along with me, and scanned around in all directions. There was nothing but craggy mountains, blowing snow, and a star-packed sky that was unlike any I had ever seen, with the muted, deep whooshing sound of mountain top winds furnishing the only sounds. The density of the Milky Way made clear where our galaxy's moniker came from, as it was so thick that it was difficult to distinguish individual stars, and it did indeed resemble a gigantic milk spill, albeit with some sparkling effects.

After a minute or so, I turned back to Merle. "Wow, Merle. Wow."

"It makes you think, doesn't it? That's why I like to come out here. And now I'm going to ask *you* to do some thinking, Ken."

"OK. What about?"

"Well, I have a little device here in my pocket that will serve as my prop, as we discuss electromagnetic energy."

This sounded very interesting, and I craned my neck over towards Merle to see what kind of amazing, futuristic device he was going to pull out of his pocket. I was surprised and confused when he pulled out an ordinary green rubber band, maybe four inches in diameter, and held it up for me to see.

"Just an ordinary rubber band," Merle said. "I see that."

"That's all we should need to come to a good understanding of electromagnetic energy."

"If you say so, Merle." I very much doubted that I would be able to understand electromagnetic energy with just a little rubber band.

"First, though, let's think a bit about what it means when we talk about the space/time reference frame of any individual observer."

"OK."

"I like to think of the space/time continuum as a series of shells, or spheres, with the observer at the center. Although one shell cannot truly be separated from another, let's imagine that somehow we can just look at one shell, or sphere, of the space/time continuum, that is presently 186,000 miles away from the observer in all directions, and traveling towards that observer at 186,000 miles a second, like a giant imploding ball.

There are also portions of space/time, at that same distance, that are traveling in a vast--actually infinite-- range of velocities, as well. But we are only concerned with that portion of the continuum traveling at 186,000 miles per second, or the ratio of space to time, relative to the observer. The key here is that the space/time continuum interacts with any single observer at that very velocity- 186,000 miles per second. That's how it works. That's why 186,000 miles per second is the active ratio of space to time, anywhere in the universe, for any observer."

"OK."

"In one second, that portion of space/time will be upon the observer. In another second, it will be 186,000 miles away from the observer again, as an expanding sphere. In ten seconds, it will be 1,860,000 miles away. In an hour, it will be almost 670 million miles away. And the observer can point to that receding shell, 670 million miles away in all directions, or 1.34 billion miles in diameter, as being his or her slice of the space/time continuum, from one hour prior. But there is another portion of space/time, 670 million miles away, and *approaching* the observer at 186,000 miles per second, that will intersect with that observer in exactly one hour."

"Damn!"

"Yes, pretty fast, right? It's all relative, though. But that's how the space/time continuum works. The continuum operates in all directions, at an infinite range of velocities. Therefore, for any single observer in the universe, there is a sphere of space/time that acts as the reference frame of the observer, moving towards him, or traveling away from him, at the ratio of space to time in the universe. And behind that shell is an endless stream of shells-- a continuum of shells, traveling at the same ratio."

"I can see that."

"Good, good. Any object in the universe that is traveling towards you, or away from you, *slower* than your shell of space/time, is part of your space/time reference frame. Any object in the universe that is traveling *faster* than your shell is in an entirely different perceptual dimension of space/time; meaning that you cannot see it, or measure it, or weigh it."

"OK."

"Now, you may be wondering how electromagnetic energy fits into all this."

"Yes, I was, actually."

"Well, that's where the rubber band comes into play." Merle held the rubber band out towards me, between his thumb and index finger. "This rubber band represents a photon."

"So a photon is a loop?"

"Actually, yes. The photon is a loop, indeed. Actually there are electric and magnetic components, each traveling perpendicular to each other, but for our demonstration purposes, this single loop of a rubber band will represent a loop of electromagnetic energy."

"I thought that a photon is a wave."

"Well, that is almost true."

"Or is it a particle?"

"Well, if you consider a loop to be a particle, then a photon is a particle."

"I've always been taught that the photon is *both* a particle and a wave!"

"Well, it's not, really. But let's move on with the demonstration."

"All right." I was highly skeptical of the demonstration, at that point. That just demonstrates how stubborn people can be when faced with truths that contradict their own "knowledge". I suppose that is exactly what Mark Twain was talking about.

"Imagine a light source, 670 million miles away, pointed towards us, and motionless in comparison with us. So it is completely in the exact same frame of reference that we are."

"OK."

"It releases a photon in our direction."

"OK."

"Now, before we get into the rubber band, and discover exactly how photons move through the space/time continuum, let's step back a bit and think about a wheel on your bicycle."

Maybe what impressed me the most about Merle was his ability to use the simplest props to explain the most complex manifestations of the universe. Whether it was a rubber band, the wheel of a bicycle, ducks bobbing on a lake, or breakfast food, he came up with some amazingly simple analogies that truly helped clarify new ways of thinking about things.

Merle continued with his bicycle analogy. "Let's look at what is happening with a bicycle wheel as you ride your bike at ten miles per hour."

"OK."

"How fast is your wheel moving, overall, relative to the ground which is typically your frame of reference on earth, if the bike is traveling at ten miles per hour?"

"Ten miles per hour, right?"

"Right, when we're talking about the wheel as a whole, or maybe if we're just talking about the axle of the wheel. But how fast is the top portion of the wheel moving?"

"Ten miles per hour, right?"

"No. Think about it. The bike is moving at ten miles per hour. So the axis of the wheel is moving at ten miles per hour, relative to the street. But the top portion is rolling in a forward direction, at just about twice the velocity of the axis, or an angular velocity of just about twenty miles per hour."

"OK." I didn't really understand the concept, at that point, but after I thought about it a bit, it was sort of obvious.

"And how fast is the bottom portion of the wheel traveling, then, relative to the ground, in terms of angular velocity?"

"Twenty miles per hour?"

Merle just looked at me for a moment. "No, Ken. Is the bottom portion of the wheel rolling in a forwards direction, relative to the axle?"

I had to think about it. "No, it's not. It's actually rolling in a backwards motion." I had never really thought of it like that before, and it sounded weird.

"That's right. The very bottom portion of the wheel is actually traveling backwards, relative to the axle, at an angular velocity of just about ten miles per hour. That counters the overall forward wheel motion of ten miles per hour. So the bottom of the wheel, which is in contact with the pavement, is virtually at a standstill, relative to the ground, at any given moment. That is how a wheel maintains traction with the road, even if it is turning quite rapidly."

Those were some weird ideas, to me, especially considering we were talking about such an ordinary, commonplace thing, which I had never really thought about much, in the past.

"So the uppermost point of the wheel is traveling at twenty miles per hour, the bottommost portion is at a virtual standstill, and various points in between are traveling anywhere from zero to twenty miles per hour, relative to the reference frame, represented by the ground. The leading edge of the wheel is traveling perpendicular to the direction of movement, with no angular velocity either forward or backward, so that portion of the wheel is traveling at the same velocity as the wheel itself, relative to the ground-- 10 miles per hour. And that is very much how a loop of electromagnetic energy travels."

# CHAPTER 24

I took a deep breath and sat back in my chair a bit to think about what I had just learned about a simple bicycle wheel. I peered up at the magnificent bounty of stars above, and a prominent meteor caught my attention as it blazed past.

Meanwhile, Merle continued with his lesson. "So let's go back to that photon that has just been released from a light source which is 670 million miles away, and completely motionless, relative to us."

"So in one hour, that photon reaches us."

"That's right. Anyhow, here is the secret to how a photon loop travels, Ken. A photon is a loop of energy that is orientated in such a way that it allows the space/time continuum to carry it along without any real impedance. Think of it as being like a rubber band, or a bicycle wheel, that is absolutely perpendicular to the road, which in this case is the continuum, which is itself a form of moving energy, like a fast, rippled conveyer belt, as opposed to a motionless road. A photon is not only propelled along at $c$, by the continuum, but it also rotates at $c$, as well, as it travels through space, as if it were actually rolling on a motionless road, at that velocity.

"As the photon is released, it is propelled along by the portion of the space/time continuum which defines the reference frame for the light source. In other words, the photon loop is propelled along by the space/time continuum, at the ratio of space to time relative to the light source. Merle held up the rubber band, in a loop shape, and moved it forward, in a rolling motion. As he held it up in the air, backlit by the bright, broad band of the Milky Way, I could imagine the photon rolling along with the space/time continuum, as it raced towards Earth from a light source that was 670 million miles away.

That gave me an idea. "So, Merle, the top portion of a photon is rotating at just about twice the ratio of space to time, relative to the reference frame of the observer and the light source which shares the same frame? And the bottom portion of a photon, at any given moment, is at a virtual standstill?"

"Yes, that's right, Ken! And the leading edge is moving at the exact ratio of space to time, relative to the light source. Not much different than how your bicycle wheel moves along the pavement."

"So how does the photon interact with me, then? How do I see it?"

"That is a great question; a very pertinent question, Ken. You, quite simply, see the portion of the loop that is moving at the ratio of space to time, relative to you."

"So... my incoming slice of space/time is moving at the ratio of space to time in the universe, relative to me. So I am able to perceive the portion of the photon loop traveling at that speed—the leading edge, in the case of our example."

"That's right! You measure the speed of the photon at 186,000 miles per second, based on the velocity of the leading edge, but what you are actually measuring is the speed of the space/time continuum that interacts with you. And the portion of the loop that you see is only that single point that is traveling at 186,000 miles per hour, relative to you. So you only see it as a point; you don't see the entire loop. Then, the energy of the entire loop is transferred to an electron within a photoreceptor cell in your eye, through its intersection at that single point."

"Wow. That's incredible!" I was absolutely blown away by the concept of the photon being a rolling loop. "So what happens if the light source is moving away from me?"

"That's another great question, Ken! Another great question! So let's say the light source is moving away from you at 100,000 miles per second, as it releases the photon. The photon loop travels away from the light source at 186,000 miles per second, relative to the light source. Therefore the loop of energy is actually approaching you at only 86,000 miles per second."

"But that can't be right! Light always travels at 186,000 miles per second!"

"One would think, right? But that's not quite what's happening when we 'measure' the 'speed of light'. As the photon loop in our example approaches, at 86,000 miles per second, relative to us, are any portions of the photon loop traveling at 186,000 miles per second, relative to us?"

I thought about that one for a bit. "Well, the photon is still *rolling* at 186,000 miles per second, I assume- just traveling more slowly, from our perspective."

"Yes."

"OK, so photons always roll at *c*, relative to their own axis. So the top of the photon is traveling at 186,000 miles per second, plus the 86,000 miles per second the overall loop is traveling at."

"So the top of the photon is traveling at 272,000 miles per second, as it arrives," Merle said helpfully. "That is more than fast enough. The bottom portion of the loop, in this case, is actually rolling away from you at 100,000 miles per second, matching the movement of the light source."

"True enough," I said. "But there is a portion of the loop, part-way between the top portion and the leading edge, traveling at 186,000 miles per hour, relative to me."

"Yes."

"And as it rides in on me, that portion of the photon is what I see."

"Yes."

"And I measure the photon as traveling at the ratio of space to time I absorbed it at, when in actuality the photon is traveling only at 86,000 miles per hour. What I am actually measuring is the slice of the space/time continuum which is incoming at 186,000 miles per second."

"Yes! Similarly, if the light source was *approaching* you at 100,000 miles per second, there would be a portion, part-way between the leading edge and the *bottom* portion, traveling at 186,000 miles per second, relative to you."

"So, no matter what speed a photon is traveling at, there is always a portion that is traveling at the ratio of space to time, relative to me, and *that* is what I am actually measuring the velocity of."

"Well, not exactly," Merle said. "What if the light source is traveling away from you, or towards you, at a velocity *greater* than the ratio of space to time, relative to you?"

I gave that some thought. "Then there wouldn't be any portion of the photon traveling at my own ratio of space to time. So I would not be able to see it."

"Isn't that what I was saying," asked Merle, "when I talked about the length contraction transformation? If something is traveling at a velocity greater than the ratio of space to time, it cannot be seen. Now you can understand that from the physical mechanism of a photon, as well."

"Wow. You're right!" I was amazed by this observation. "Either by the Lorentz transformation, or by the physical mechanism of the photon, you can't see a light source that is traveling at speeds greater than 186,000 miles per second, relative to the observer."

"That's right!"

"Wow, Merle, I just realized what the Lorentz transformations are, really! The space/time continuum allows us to perceive other objects, relative to ourselves—in terms of being able to see things and measure their length, in terms of being able to feel things and measure their mass, and in terms of being able to spend time together and measure the passage of time. The Lorentz transformations describe how the continuum shifts our perceptions of a traveler, at relativistic velocities: Mass appears to increase towards the infinite, length appears to contract towards disappearance, and clocks appear to slow down towards meaninglessness. The Lorentz transformations are just showing us how the space/time continuum works, in terms of allowing us to perceive portions of our universe in a frame-shifted manner, but only up to a certain velocity. The transformations define the physical reference frame of any single observer—the four dimensional frame of the observer."

"You can also see the implications for the First and Second Postulates."

"I can?"

"Sure. Now we see that the First Postulate and the Second Postulate are both offshoots of the transformations, really. Although the Second Postulate doesn't really apply, once a traveler exceeds $c$."

"That doesn't stop us from accelerating to any possible velocity we can think of, though, does it? We can go as fast as we want!"

"Yes, of course we can go as fast as we want. We sure can! I've never heard of any big walls out there, in space. We're just in a different physical dimension, if we go fast enough. Or, I should say, if we shift our perceptual frame in a certain direction."

"But wait, Merle!" I just had a disturbing thought. "I can see how the photon fits into all this—into the hyper-dimensional space/time continuum, and the transformations, and all that—but how does a photon appear to travel as a wave?"

"Ah," said Merle. "Yet another fantastic question!" He held up the rubber band again, and began rolling it through the air. "You see, Ken, not only does the photon roll, but it also ripples, as it rolls." He tried, fairly unsuccessfully, to make the rubber band smoothly ripple as he held it in a very rough approximation of a loop. "That's why I am using a flexible rubber band in this demonstration, although it's hard to make it ripple the way I'd like it to. Anyway, a high energy photon rides the space time continuum with a lot of power, you might say, so it is more significantly rippled, as it rolls, than a lower energy photon. Either way, the entire rubber band—or photon—ripples, as it rolls."

"The high energy photon has a short wavelength," I said.

"That's right. The rippling comes fast and strong with a high energy photon. High crests, short wavelength."

"And a low energy photon?"

"Well, that ripples more gently, with longer ripples, and lower crests."

"Long wavelength with the low energy photon," I said.

"That's right. Let's look at our original example, where we end up seeing the leading edge of the photon."

"OK."

"In a high energy photon, the leading edge is bouncing up and down, quite rapidly, due to the rippling. If you plotted the course of the leading edge, as it bounces and rolls through the continuum, you'd see that it manifests itself in a wave."

"Wow! I do see that." As Merle rolled and bounced the leading edge of the rubber band up and down, the visual was obvious. Traveling forward while moving up and down pretty much describes a wave, all right.

"Either way, whether it's a high-energy photon, or if it's a low-energy photon, the observer will interact with a portion that is rolling along with the ratio of space to time, relative to him or herself, and it will be perceived as traveling at the ratio of space to time, in a wave. Even though it's just a rippled, rolling loop of energy, that is quite possibly *not* actually traveling at the ratio of space to time, relative to the observer." Merle stopped bouncing the rubber band for a moment, as if to reinforce what he was saying.

"Although the photon is really a trans-dimensional object, due to its velocity range, the diameter is always the same, whether high-energy or low-energy. That's why Planck's constant, as it's called on Earth, is measurable as the constant ratio that it is." The Planck constant was another concept that had always interested me, and it was pretty neat that Merle's model of a photon fit it perfectly.

Merle continued. "Another point to consider is that while the loop as a whole is rippled, the bottom portion is much more heavily rippled than the top of the loop."

I was a little confused by the concept of the rippling. "I'm not sure I see why the photon ripples, Merle."

"Well, think of it this way, then. The photon is rolling forward, along with the space/time continuum. The top portion of the loop is moving in the same direction as space/time, so it's a smooth ride, you might say. Also, the top portion has to move forward very rapidly to keep up with the main axis of the photon, so to speak. So there's no reason, and no time, for it to be bouncing up and down very much. It only has a slight ripple, perhaps."

"OK. That makes sense, I guess."

"The bottom portion, on the other hand, is going backwards, against the flow of space/time, to some extent. By the same analogy, the ride is not as smooth on the bottom as it is on top- you might say it is fighting against the stream. At any rate, the bottom portion, with its backward motion, naturally ripples much more heavily. That is how it manages to keep in sync, velocity-wise, with the much straighter section at the top." Merle crinkled the rubber band at the bottom while keeping it much more directly horizontal at the top.

"That is why, if a light source is moving away from an observer, and the intersecting point is more toward the top of the photon loop— which is bouncing less than the leading edge-- the wavelength gets longer, or is red-shifted, you might say. Conversely, if a light source is moving towards the observer-"

"The wavelength shortens, as the intersecting point moves towards the bottom of the loop, which is bouncing more, and therefore blue-shifted," I said.

"Yes!"

"So the bouncing is really from the interaction with the space/time continuum?"

"Yes. Actually, it's more accurate to that the photon rides the space/time continuum, which travels not only in an infinite range of velocities, but also in an infinite range of wavelengths. The space/time continuum itself is the wave, you see. The photon just finds its place within. The quantum loops of energy we call photons interact with the space/time continuum in the range of wavelengths that we have observed and documented. Whatever wavelength is needed, based on the actual energetic qualities of the photon, is delivered by the infinitely variable space/time continuum." At that, Merle stopped and watched for my reaction.

I thought about those words. "The space/time continuum itself is the wave, then. Not the photon itself?"

"That's right."

"Oh. My. Gosh! Merle! That is a huge concept!"

"Yes, and not only does the continuum influence small quantum units of mass- energy, like photons, but it also influences larger, more massive units, like you, or me, or Planet Earth, if only through the passage of time. Time itself is a wave phenomenon, really, but the effect is vanishingly subtle, on our level.

"Perhaps you can begin to understand, Ken, what I meant by saying that the space/time continuum itself represents energy, of a different sort. You see how it physically interacts with electromagnetic energy, and how it travels in a wave. Those aspects of the space/time continuum describe a direct, energetic manifestation. One might even think about the passage of time itself, another effect of the space/time continuum, as a similar energetic manifestation upon mass/energy, exerting change upon it. And, obviously, gravitation is the involuntary movement of mass/energy through the continuum, in a particular direction, which is an energetic action, in itself."

So that is how Merle taught me how a loop of electromagnetic energy interacts with the space/time continuum, always appearing to travel at $c$, and always appearing to travel in a wave, while manifesting itself as a particle. He used a rubber band as his prop, my imagined bicycle wheel as a secondary example, and the Milky Way galaxy and Himalayan mountain range as the impressive backdrop. Another important thing I gained perspective on that day was that the space/time continuum is actually a *thing* in the universe—an important moving field of energy of sorts, which affects every quantum manifestation, one way or another. The continuum is not just some abstract, irrelevant, inconceivable concept, but an understandable force, with infinite range, that unfailingly follows a set of rules.

I was struck with a thought that actually disturbed me. "So are you saying, Merle, that the light from a distant galaxy is actually traveling to Earth at velocities below the ratio of space to time?"

"Absolutely."

"So that must mean that those distant galaxies are actually much farther away than we realize! And the visible universe must be far older than what we have been assuming!"

Merle laughed. "On the other hand, these now-distant galaxies were much, much closer to us when they released the photon than your astronomers believe. So maybe they are not quite as much older as you might think. Then again, maybe they are. It gets sort of complicated." He winked. "Remember, I can't give you *all* the answers. What would be the fun of that?

"See the sky up there?" Merle asked me. "That is just the portion of the universe that is traveling at velocities below the ratio of space to time, relative to us. But the portion that is traveling beyond our perceptual dimension of space/time is literally endless. In other words, our perceptual frame of space/time is not even a grain of sand, compared to all the matter in the universe. Nowhere even close to a grain of sand. We're just a grain of sand on an infinite beach."

I'll never, ever forget the realization I had, at that moment. There we were, sitting on a desolate, snow covered mountaintop at night, somewhere in the vast, seemingly endless expanse of the Himalayas, our surroundings brightly illuminated by a brilliant blanket of hundreds of millions of blazing stars strewn across the boundless sphere of the sky. Just one single star was nearly incomprehensibly massive, to my previous way of thinking, but suddenly I realized that all of it represented not even the tiniest speck, in the grand scheme of the multi-dimensional universe. All the grandeur, all the vastness, the totality of all the distant galaxies that have ever been imaged by Earthly telescopes, was a mere dust mote of near-insignificance, compared to the unseen portions of the universe. At this point, I was still some ways from fully understanding the real nature of the "greater universe," as Merle put it, but even so, I was overwhelmed by the thought of what I *did* already know. I was getting quite emotional, the realization was so intense.

Merle put his hand on my shoulder. "It's OK, Ken. I know it can be overwhelming. I've put a lot on your plate here, today."

"No, no," I lied. "It's fine."

"Well, I think it's probably just about time for us to call it a day. You still have to go to work this afternoon, don't you?"

That comment shot through me like a lightning bolt. "Crap, Merle! I've got to get back! I *do* have to work today! I was supposed to be at The Enterprise at 1:00!"

"Not a problem, Ken. Not a problem. We've got plenty of time."

"You mean it's not even 1:00 yet?"

"No. It's only about 10:30 in the morning, still, back in Rundle Heights."

"Wow! How in the heck is that possible?"

"You left the house very early, this morning."

"I guess I must have. But it seems like we've been out and about for a full day, already."

Merle laughed. "Not quite! Less than four hours."

"Less than four hours? Damn! How did we do all this in less than four hours?"

"Moving fast, I guess. Speaking of which, we should probably get going."

"Yes."

And with that, Merle touched the screen, our spaceship reappeared around us, and we lifted back off the surface of the mountain. It all took about three seconds, and we were in the air. Merle just continued speaking, matter-of-factly, and no differently than as when we first pulled out onto the street that morning. "Do you realize, Ken, you are the first person, in the history of the Earth, to understand the Lorentz transformations as you do, and understand the space/time continuum and electromagnetic energy as you do?"

I didn't really know how to respond to that. I was pretty freaked out by that thought. "No need to answer that, Ken. I'm happy for you, though, really happy. You worked for it, too, which is great. Now let's get you back home in time for your job."

Merle was not kidding. He had me back home by 10:45 a.m., in fact, which I had a hard time believing. I was exhausted from the events of the morning, so by 11:00 I was asleep on the couch. I napped for almost an hour, and then I still had some time to sit around the house, before leaving for work, and to think about what had happened that day. I wrote down everything I could recall, which helped me out later, when I decided to write down my experiences. After I came home from work that evening, I made some dinner, and I wrote some more, until bedtime. By now, I was totally hooked, regarding this new way of thinking about the universe.

# CHAPTER 25

That night, I dreamed of flying through space. I can't remember exactly what it was all about, but I know I was gliding through an endless field of stars. They were so thickly spread that it was like swimming through a dense and limitless school of shiny little fish in the ocean. I had to sweep the stars out of the way, with my hands, as I flew along, to avoid getting them in my eyes. I wasn't in a spaceship; it was just me, in my pajamas, flying through the universe, all alone. I wasn't scared, either. In fact, it seemed very reassuring and peaceful. That's all I really remember about it.

When I woke up, I was in a bit of a stupor, still drained and dazed by my dream. I remember that I opened my eyes, bolted straight upright in bed, and immediately thought about Merle. I realized I had no way to get in contact with him. We hadn't discussed any of that, the day before. I jumped up, showered, and quickly made some coffee and gulped a cup down. I assumed that I had to go to the forest preserve again to meet him, and it was already getting late. As I was in the kitchen getting ready to leave, I briefly heard the now-familiar droning sound again—even though I was still in the house. I glanced out the window, and I noticed that somebody was on the front porch. I looked out the blinds for a closer look, and I saw that it was Merle! I dashed over to the door and threw it open. "Merle! How long have you been here?"

Merle looked at me and raised a finger to his lips. "Ssshhh. You're scaring the cat."

"Cat? What cat?" I looked down and saw a little black and white cat, backing away from me, apparently getting ready to take off, down the stairs.

Merle called out to it. "It's okay, little kitty. It's okay. You're still hungry, aren't you? Come here... I have your food." Merle reached down and picked up a little dish with dry cat food in it, and moved it closer to where the cat stood, poised to run away. The little cat thought about it, and then cautiously moved back towards the dish. Merle motioned to me to close the door, quietly, and move away a little to give his little friend some space. The cat furtively walked over to the dish, and began to eat. Or, I guess more accurately, resumed its eating. "She sure is hungry," Merle said.

I noticed a water dish, filled with water, next to the food dish, and an entire bag of cat food in the corner of the porch. I was confused as to how it all got there. "Where in the heck did the food and all this come from?"

"Clotro brought them over. This little kitty needs some attention."

"Clotro? Where is she?"

"She's gone now."

"Darn, I missed her!" I still wasn't used to how quickly things happened with this group. "Whose cat is it?"

"Clotro says that she's a little stray cat that your neighbor has been feeding for the last few weeks. But your neighbor hasn't been around the last couple of days. I guess she's in the hospital."

"Really? Neddie is in the hospital?" Neddie was the elderly lady that lived next door. I hadn't noticed that she hadn't been around. I had been a little busy, the last couple of days. I didn't know her very well, but Walter had told me that she was very nice.

"Yes, apparently she had a fall."

"That's too bad," I said. "I hope she's going to be O.K."

"It's nothing too serious, is my understanding," said Merle. "She just needs some time."

We both stood there and watched the cat. It was still gobbling food, ravenously. At the same time, out of the corner of my eye, I noticed that a car had parked along the curb, behind Merle's car, and a young woman came out. She came up the sidewalk, and I realized that she was approaching us.

"Hello!" she said, as she approached. "Is that your cat?"

"No," said Merle. "She's very hungry, though, so we thought we'd better feed her."

By now she was on the walkway, in front of the porch, watching the cat eat. "Oh my gosh! You guys are so sweet! This must be 'L.C.'!"

"Her name is 'L.C.'?" Merle asked.

"I'm not sure if it's a 'him' or a 'her', she said. But my Grandma Neddie named it 'L.C.' It stands for 'Little Cutie.'"

"What a great name!" said Merle. He looked over at me, and then at the woman. "Hi," said Merle. "I'm Merle, and this is my friend, Ken. He is staying here. I just stopped by."

"Hello!" said the woman. She made her way up the steps, slowly, so as not to scare L.C. "I'm Kim." She extended her hand, and Merle and I both shook hands with her.

"Nice to meet you," she said. I noticed that her hand was very warm, soft and smooth. My hand was rough and chapped from garden work, I was pretty sure.

I immediately liked this woman, Kim. She appeared to be about my age, and she seemed to be rather friendly and nice. She was neatly but not ostentatiously dressed, with sensible shoes, a sharp but simple hairstyle, and she wasn't overly slathered in makeup or in-your-face attitude. "Probably already has a boyfriend," was the thought that popped into my mind.

"My grandma has been feeding this little guy for several weeks," she said, "but four or five days ago, she tripped over a pair of shoes on the floor, and hurt herself. Now she has some complications, and she's still in the hospital. She's been so worried about L.C., here, that I promised I'd come over and see if I could help. Wait until she finds out that you were already feeding him, when I got here!"

At that, the cat stepped back from its food dish, and moved over to the water dish, and lapped at the water with its tongue, for more than a minute, I'd guess. Then it stepped back again, lifted one of its front paws to its mouth, and gave it a few good licks. Then the cat groomed itself, behind the ears, with its wet paw. The cat sat down next to

Merle, apparently satisfied. Merle bent down, reached out a hand, and said "here kitty, here kitty!" Merle held the position, motionless, and the cat looked up and walked over to him. L.C. arched up her back like cats do, while fluffing out her tail, and rubbed up against Merle's leg. Merle reached right down and gave his new friend a nice little rub, behind the ears.

"Looks like you have a friend, Merle," said Kim. "She's a sweetie, all right," said Merle.

"She?" I asked. "How do you know it's a 'she'?"

"I have a pretty good hunch."

I figured that Merle was probably quite certain about his "hunch". "Also", said Merle, "she has a wound."

Kim had been leaning back against the railing, watching L.C., and she bolted upright at that. "What? What kind of wound?"

"It's not too bad," said Merle. "There's a small wound on her right shoulder. She's licked the fur off- see right there?" he said, pointing. "It doesn't look too serious. I don't see a deep puncture, so I don't think it was an animal attack. It looks like she might have taken a tumble, out of a tree or something."

Kim looked horrified. "Oh, no! What am I going to tell my grandma? She's already worried sick about this cat!"

"Maybe we should take her to the veterinarian", I said.

"That's what my grandma was talking about, before she went into the hospital," Kim said. "She was thinking about getting her checked out at the vet, and then maybe seeing if she might like to stay in her house. She hasn't had a pet since her dog died, a few years ago, and she loves this cat."

"Maybe we can take her to the vet, and you can take her into your house until Neddie gets out of the hospital," I offered.

"I'd love to, but I can't. My dad is very allergic to cats, and he already specifically told me not to bring the cat back home with me."

Merle turned to me. "How about you, Ken? Do you think Walter would mind if you brought in a cat for a few days?"

I had never had a cat, or even known a cat. I considered myself to be a "dog person", but definitely not a "cat person". I was pretty sure that I wasn't going to be able to deal with a cat. I was pretty sure that I wouldn't really like a cat. "I don't--" I started to say, but before I could get any more words out, I was interrupted by a sudden crashing noise, coming from inside the house.

"What was that?" Merle asked me. "You'd better go check."

"I'll be right back," I said. I opened the door and went back into the house. I could see a frying pan lying on the floor, which was weird, since I didn't remember taking the pan out. As I bent down to pick it up, I heard the door shut behind me. As I spun around, the room began to visually melt away before me, transforming into a completely bizarre, shimmering, light blue haze. My first thought was that I was having a stroke, or something. Disoriented by the sudden transformation of the kitchen into a hazy limbo, I turned around, and I was startled to see Atropha standing there, apparently floating amidst the fog of blue.

"Are you a jackass?" she asked me.

"Excuse me?" I was struck by how surreal the moment was, and I had to check myself to see if I was dreaming. I still wasn't sure.

"I said, 'are you a jackass?'" she repeated. "What do you mean by that?"

"You were about to say that you don't think you can take that cat in, weren't you?"

"I don't really like cats. I wouldn't know what to do. I don't think it would work out."

"*Three excuses!*" Atropha was getting fired up. "You don't think it would work out! Well, *I'm* wondering if it's going to work out, also!"

At this point, I was very confused, and more than a little frightened, I have to admit. "How did you get in here?" I asked her.

"Yes, you are a jackass," she said. Alarmingly, she took a step towards me, whipped something out of her side pocket, and waggled it at me. It looked like some sort of sparkly, futuristic weapon, and I instinctively jumped back.

"What is that?!" I was in danger, I assumed, and I was wondering if there was any chance that I could make a break for the back door. But I had no idea where the door even *was*. In fact, I couldn't make out a single detail of what used to be the kitchen, amidst the bizarre sparkling blue. Now that I had a closer look at her weapon, though, I could see that it resembled a pair of garden shears- a small pair, the kind you can use with one hand. But these "shears", which looked like something out of the 28th century, looked particularly menacing and dangerous, and I thought I could see tiny sparks flashing of the sharp-looking ends of the "blades". I very seriously doubted they were garden shears, and I backed away a bit, even though I couldn't see a thing other than the fog, and Atropha. She ignored my question and continued to waggle the weapon at me.

Right at that moment, a loud cracking thump rang out, and I spun around in the other direction. Standing there, or more accurately, hovering there in the shimmering blue blanket of light, was Latsis. Things were definitely getting stranger, by the moment.

Latsis stepped towards Atropha. Latsis held her staff upright in her right hand, her arm extended outwards towards Atropha. She waved her staff with emphasis, as she spoke. "And what do you think you're doing, Atropha?"

"I'm saving this jackass from himself, that's what I'm doing! He's weak! He's selfish, and he only thinks of himself!" She waggled her weapon at me some more, for emphasis, apparently. "He was a completely ridiculous choice, as I said from the beginning! The boy won't even offer assistance to a small animal! He 'doesn't think it would work out,'" she said, clearly mocking me in a whining tone of voice. "He's all for himself!"

"You know that it's too late for us to stop now," Latsis fired back. "We cannot just give up on the boy. You should know that."

"Bah!" shouted Atropha, and she stomped her foot into the fog, apparently striking the floor, from the sound of it. "But is there time? I doubt it!" And with that, she glared at me, fiercely. As I took another step back, she continued to glare, but I noticed that she began to slowly fade away, out of sight, like the giant triangular ship had done, out in the forest. The last things I saw, as she faded out like some demented Cheshire cat, were her fierce scowl, and a glint from her frightening "garden shears".

I was left there, alone with Latsis. I was very much shocked by the turn of events. "Ken," said Latsis.

I was afraid, and too embarrassed, to even look at her. "What?"

"Look at me, Ken."

I looked at her. "What?"

Latsis looked at me and smiled. "We still have plenty of time, Ken. You're in this for the long haul, you know. Everything doesn't have to happen in one day."

"It sure *seems* like everything has been happening in one day."

"Ken," said Latsis, again. She walked towards me—floated towards me, is more accurate—in a completely non-menacing way. She stood directly in front of me, and before I could think of something to say, she put her arm around me. It was the most wonderful feeling. "It's going to be all right, Ken. I will help you." She stepped back and smiled at me. "Now, were you serious about not wanting to offer your assistance to that poor little animal?"

Well, I had softened somewhat on that stance, I suppose, after hearing Atropha's "persuasive" tirade on the subject. "I don't know. I don't know if Walter would want me to have a cat in here."

"I don't think he would mind, Ken. Why don't you call him and ask him, yourself?"

"I can't call him right now. I have to get back out to the porch. They'll be wondering what has taken me so long."

"No, they won't," said Latsis. "Why don't you call Walter and ask?"

"Can I use my cellphone? It's on the kitchen table. Wherever that is."

Latsis already had the phone in her hand, and she reached out to hand it to me. "No pictures!" she said.

"I know, I know. I haven't been taking any."

"I know. That is good. OK, you can make the call."

I made the call. Walter liked the idea, but he was surprised that I was going to bring a cat in. "I thought you didn't like cats," he said.

I told him that I had a change of heart, and I was going to give it a try. "Besides, Neddie really needs the help", I added.

After Walter and I ended our conversation, and said good-bye, I pressed the "end" button on my phone. That's when I realized that the fog in the room had lifted, and Latsis was nowhere to be seen. Thankfully, Atropha was still gone, as well. I put the phone back down on the kitchen table, and headed back out to the front porch, more than a little shaken up by the bizarre events that had just unfolded.

I opened up the door to the porch, and Merle and Kim were still engrossed in the cat. "It was just a frying pan that fell," I said. "Also, I called my cousin Walter, and he said it would be OK if I took in L.C. for a while."

Kim looked at me, surprised. "You already called your cousin? That sure was fast! You've only been gone about a minute! Wow, that is really super nice of you! My grandma Neddie is going to be so happy that we're going to help her little kitty!" She sort of took a half-step towards me, and for a moment, I thought that Kim was going to give me a hug. I think she quickly realized that we had just met a few minutes ago, and maybe it was a little soon for a hug.

I was a little taken aback by her saying that I had only been gone one minute. I thought it had been *ten* minutes, at least. I suspected the blue fog had a lot to do with it. Later, Merle told me that I had actually experienced a "reverse time dilation", as performed by the time-savers. He told me that the engineers back on the base ship-- the triangular ship—were all blown away by the maneuver, and they had no idea how the time savers pulled it off. I guess the whole ship was buzzing about it, afterward, from what Merle said.

As I stepped out onto the porch, I heard a car door shut on the street, and I looked out to see that my parents were walking up the sidewalk towards the house. I panicked, big time, about Merle being there.

My dad called out to me, as they were walking up. "Well, Ken, we didn't expect to see you up so early!"

My mom interjected, "we were on the way to the Garden Center, and we saw your friends out on the porch, so we thought we'd stop by and say hello. If I knew you'd be up and about already, I would have brought over some of the pot roast I made last night!"

"That's OK, mom," I said, while desperately trying to think of something to say, to explain my situation. Merle bailed me out, on that one.

"We were going to go out for breakfast, anyhow," said Merle, "to The Enterprise."

"Oh, going out to The Enterprise! That's nice!" said my mom. By that point, they were walking up the steps, and L.C. got frightened and jumped off the side of the porch, bounding over Neddie's short, decorative fence and into her backyard.

"Whose cat is that?" asked my dad.

"That's L.C.", said Kim. "My grandma Neddie had been taking care of her, but my grandma is in the hospital, right now. She broke her foot, and had some complications."

Kim introduced herself to my parents, and, as it turned out, my mom had known Kim's grandma Neddie for many years. Neddie was my mom's teacher, in the fifth grade. My parents were very sorry to hear about Neddie's foot, needless to say.

Then Merle introduced himself, while I secretly cringed in horror. "So where do you know Ken from?" my dad asked Merle.

Merle surprised me with his answer. "We go to college together," Merle fibbed. "I was in Ken's freshman Physics class, with Professor Thomas."

"Professor Thomas, huh?" my dad said. "He was a tough one, wasn't he?" Merle agreed. "Yes, he was a little tough. Ken was the smartest kid in the class, though, I thought."

"Ken was?" my dad said. "I thought you got a 'C' in that class, didn't you, Ken?"

"Yes, I did, but…"

Merle interrupted me before I could get a sentence out. "You might have gotten a 'C', but still you always asked the best questions."

My dad laughed at that. "Hah! Hard to believe you got a 'C', then, Ken!"

I laughed, very uncomfortably. Luckily I was rescued by Kim, who rather shocked me by announcing that she, also, attends the same college, studying business, and was due to graduate in the same year that I was. She had gone to a local community college for the first two years, and was finishing up the final two years of her degree at my school. She said that last year was sort of difficult because she didn't know too many people there, other than the girl she was sharing an apartment with. We all talked about what a coincidence it was, and my mind started to race a bit, thinking about the possibilities. Luckily, nobody asked Merle too many questions about *his* college life. I have a feeling, though, that he would have been ready with some very believable answers.

Finally, Kim wrapped up the conversation. "Well, I have to get going. I'm on my way to work. I'm doing an apprenticeship at an accountant's office this summer, and since I couldn't make it in yesterday, I have to put in a few hours today."

Kim and I agreed to meet back at my place at noon. "Hopefully L.C. will still be around, and we can take her to the vet," I said.

"I'll pay for it," she said, "if you'll take him in, afterward, until my grandma gets out of the hospital."

"Deal."

"Great! I'll see you at noon, then!"

"OK. See you then!"

"OK. Bye, then! Nice meeting you, Mr and Mrs Sylvanewski! You too, Merle!" And with that, Kim went back to her car, and drove away.

My dad looked at me with one of those "You dog!" smiles. "You sure do move fast, don't you, Ken?"

"She seems like a very nice girl," my mom offered. "She is very nice," Merle agreed.

"She probably has a boyfriend," I said. At that point in my life, I was in the midst of a long spell of either no dating, or catastrophic dating. My romantic efforts had become a running joke amongst family and friends, and I was well on my way to becoming seriously jaded about the whole thing. I guess I had gotten used to ruling out relationships at first glance, especially with appealing candidates, who seemed to always be in a relationship, already. The whole thing had gotten quite discouraging, and even at this point, with a noon date, of sorts, with a promising girl, I wasn't getting my hopes up.

# CHAPTER
# 26

My parents and Merle and I hung around on the porch for a while, and made some small talk. I was relieved that the conversation didn't stray into any uncomfortable areas. After a few minutes, my dad got a little antsy to get over to the Garden Center, and we said our goodbyes.

I was curious about Merle's plans for the morning. "So do you really want to go to The Enterprise, Merle?"

"Absolutely! I have some new areas to discuss, and I thought it'd go a little easier with some pancakes."

"Pancakes?"

"Yes."

On the way over, we listened to classical music in the car. It was the "Sonata in B Minor", by Franz Liszt.

"What an interesting man Liszt was," Merle said, as we drove along. "He was an extremely influential person. He lived a most incredible life, really; a tremendously full life. He was one of the finest performers of any kind that there's ever been, on this planet, an absolutely *mesmerizing* talent! You wouldn't believe how the ladies swooned when he played, Ken!

"You know, Liszt pretty much invented the piano recital, as we now know it. Although when he played the instrument, he was very flamboyant, and physical. He performed his music as visual entertainment, and he had the crowd eating out of the palm of his hand, believe me!

"He became quite a prolific composer after he settled down a bit later in life. He was a very charitable man, as well, and donated enormous sums of money to the less fortunate. He was quite a man!" Merle leaned forward, and drummed his fingers on the dashboard as he listened, following the music note for note with his drumming.

It was nice seeing Merle so peaceful and happy. I couldn't help smiling, myself, as I leaned back in my seat to enjoy the ride. As we got closer to the Enterprise, I noticed a hooded figure, holding a wooden staff and leaning against the wall of a building, watching us drive past. "Latsis!" I shouted out. "Merle, there's Latsis!"

"Yes," said Merle, without the slightest hint of surprise. He paused his drumming to the music just long enough to give Latsis a friendly nod, and a "thumbs up" gesture. I saw Latsis respond with a "thumbs up" of her own. "Latsis has some incredible footage of Liszt, when he was in his prime, performing. I could watch that all day long," Merle said. "Latsis is who really turned me on to Liszt."

"Is this Liszt, himself, performing, that we're listening to now?" I asked.

"This? No, it's not, unfortunately. Although I have to admit that this version, by Zimerman, must be one of the finest performances of the piece I've heard by anyone not named Franz Liszt. It's a phenomenal performance, really! If you were to hear Liszt himself performing this, though, you'd hear the difference, believe me. Something you have to see to really appreciate, though."

I didn't know much about Liszt, at the time. OK, I didn't really know anything at all about Liszt, but even so, he sounded a lot more interesting than I would have thought possible. Afterwards, I did a little research, and I could see why Latsis and Merle were so enamored with him. It was pretty wild to imagine Latsis, back in the 19th century, checking out a Liszt performance, and recording it. That was just a few weeks ago, or something, for her, probably. I wondered how many other interesting things the time savers had in their "extensive archives".

Walking into the R.H. Enterprise, the first thing I noticed was the owner, James Teakurk, back by the kitchen area, talking with one of the waitresses. "There's Mr. Teakurk," I pointed out to Merle.

"The owner, right?" said Merle.

"Right." I wondered how he knew that. "He's here pretty much all the time, except for Fridays. That's the one day he takes off."

"That must be why he was at the park yesterday, while we were there."

"Out at Streamside?"

"Yes. He was just starting his morning jog, as we were heading back to the car."

"No kidding? I wonder if he saw us?"

"I think he did," Merle said.

We were shown to our table, and the waitress, Lillian, who I knew personally from working breakfast on a few occasions, came by to take our order. I ordered eggs and toast with coffee, while Merle ordered pancakes *and* a bagel, with a glass of tomato juice. Lillian smiled when he ordered both pancakes and a bagel. "I hope you're hungry," she said. "Most people can't even finish the pancakes. They're pretty large."

"Larger than the bagel, I hope!" said Merle. "Oh yes. Quite a bit."

"Good, good. That's what I wanted. Excellent! Oh, and one more thing," said Merle. "Can you try to get me a bagel that doesn't have a very large hole in the center?"

Lillian gave him a rather funny look. "Sure. I'll try to find one with the smallest hole, if you'd like."

"Excellent! Thank you very much!"

I didn't know what Merle's angle was, asking for a bagel with a small hole. I think I may have been more surprised that he seemed to be so hungry. I had never seen Merle eat anything, yet. I was wondering if he would eat Earth food, or what. Now I guessed that he obviously did.

After a few minutes passed, Lillian came back with our drinks. Mine was a coffee, and Merle took a long sip from his tomato juice. Lillian went back to the kitchen, and Merle commented on his beverage. "I like tomato juice," he said.

"I didn't think you would eat Earth food," I said.

"Well, not too much. Just certain things. Like tomato juice."

"Like pancakes and bagels, too, you mean!"

Merle shook his head. "Not so much. I guess I could eat them, if I was starving."

Before I could ask him what the heck he meant by that, after ordering both items for his breakfast, I noticed that Mr. Teakurk was headed over to our table. "Merle, Mr. Teakurk is coming over to our table!"

"Great!" said Merle.

I wished that I felt the same way, but I was always nervous about Merle, my friend the space alien, meeting anybody that I knew well, like my parents, or the owner of the restaurant I worked at.

Mr. Teakurk walked up to our table and greeted me. "Hello, Ken! Nice to see you here! One night you don't have to work in the lounge, and you're still here, having breakfast! Now that's impressive!"

I stood up and shook Mr. Teakurk's hand. "Hi, Mr. Teakurk."

"And who is your friend?"

Merle stood up, also, to shake Mr. Teakurk's hand. "Nice to meet you, Mr. Teakurk.

Merle Akeetheran."

Akeetheran! I almost spit out my coffee, when he said that. Merle told me, afterward, that his real last name was "Yamersh", or something like that, with the accent on the "mersh". But I still always think of him as "Merle Akeetheran".

"Well, nice to meet you, Merle!" said Mr. Teakurk. "Didn't I see you boys out at Streamside Park yesterday morning?"

"Yes, that was us," Merle said.

"What were you doing, picking up garbage or something?"

By now the conversation was just between Merle and Mr. Teakurk, so I just sat there, ready to intercede if things got weird.

"Well," said Merle, "we had already played some basketball, and were headed back to the car. We were just picking up some litter on our way out."

Mr. Teakurk laughed. "Picking up some litter? Do you always do that?"

Merle seemed confused as to why Mr. Teakurk was laughing. "Well, yes. There is always litter out there, so I always collect some on the way out, to put in the garbage receptacle, where it belongs."

Mr. Teakurk stopped laughing. "That is actually a very nice thing, Merle. I like that."

"Just imagine if all the people who went to the park picked up a few pieces of litter on the way out," Merle said. "After a couple of days, there wouldn't be any litter left out there."

Mr. Teakurk put his hand up to his chin, and looked at Merle. "No, I guess there wouldn't, would there?"

"No, there wouldn't. So I just try to do my part."

"That's very admirable, actually."

"Ken, here, was mentioning how polluted the stream is, too. He was saying that we should organize a cleanup. There is a lot of junk in there- plastic, shopping carts, old tires, and all kinds of other garbage. It's a hazard."

Mr. Teakurk looked at Ken. "A cleanup of the stream, huh? I like that idea! I go out there every Friday morning for my jog, and I'm always disgusted by all the junk in that stream. I always wonder why nothing is ever done about it."

"Maybe we can do something about it," Merle said.

"I like how you boys think!" Mr Teakurk said. "Listen, I have to get back to checking on my customers, but you boys let me know if there's anything I can do to help with that cleanup." He began to turn away, and then he stopped and turned back to us. "I'll help sponsor the operation, if you'd like, Ken. You'll need sponsors. My brother-in-law manages a waste disposal company, and I'll ask him if they can cart away all the trash. I'm sure there are many dumpster loads of trash in that stream." He began to turn away again, and then stopped and came back again. "Oh, and from now on, when I go jogging out there, I'm going to collect some litter on the way back!" He gave us a big, broad smile, and patted me on the back. "OK, I have to get going.

Thanks, boys!" He walked away, whistling, and began to go table-to-table, checking on his customers, as was his customary practice.

"Nice man," said Merle.

"Yes," I said. "I guess you're really going to make sure we get that stream cleaned up, aren't you?"

"Isn't it the right thing to do?"

"I guess it is."

"You know it," Merle said, as he took another sip of his tomato juice.

Moments later, Lillian came back over, with our food. She served it, and mentioned to Merle that she found a bagel that was almost closed off in the center, with just a fairly small hole.

Merle peered at the bagel. "Perfect!" he said. "Absolutely perfect! Thank you!" Lillian was trying to hide the odd look on her face, without too much success.

"You're welcome," she said. "Is there anything else I can bring you?"

"I'm good," said Merle.

"Me too," I said.

Lillian walked away, and I began eating my eggs, while I watched to see if Merle was going to try to eat his food, or what. His bagel was on one plate, while his pancakes were on a separate plate. He began by lifting off the top half of the bagel, and placing it off to the side, on his napkin. The remaining half-bagel was on his bagel plate, cut side up. He then lifted one pancake off of his stack of three, poked a small hole in the center, and placed it, well-centered, on top of the cut side of the half-bagel. The pancake extended a couple of inches past the diameter of the bagel. Then, he took the other half-bagel, and placed it on top of the pancake, cut-side down.

"A pancake-bagel sandwich?" I asked.

Merle looked at me and laughed. "Don't think of this as food," he said. "Think of this as the center of the Milky Way galaxy."

"The center of the galaxy?" I guess I should have been prepared to segue into astrophysics, but for some reason, this took me by surprise.

"That's right. The pancake represents the accretion disc, around the central black hole of the galaxy, which is represented by the bagel. Actually, the pancake should be a lot larger, in comparison to the bagel, but I didn't want to ask the waitress for a pancake that was 50,000 feet in diameter! For our purposes, this set up will do just fine."

At the center of the Milky Way galaxy is a mid-sized black hole. Around the black hole swirls a gigantic "accretion disc", which is a flattened, circular mass of stars and other mass-energy-- the equivalent of between four and five million of our own suns, according to estimates (which may be low). This disk of cosmic debris is being gravitationally herded together for its eventual spiraling descent into the black hole.

The accretion disk features a large central bulge, where mass/energy builds up in areas closer to the axis of rotation due to the smaller diameter there, much like waves at high tide bulge up upon the increasingly shallow waters of a beach. This central bulge, alone, is 12 million light years across. Earth exists far outside of the accretion disk, out on one of the spiral arms of the galaxy. The entire galaxy, itself, is 100 million light years across. Eventually, in billions of years, perhaps, our portion of the galaxy will also be drawn into the accretion disk, and eventually into the black hole itself.

"How is that bagel supposed to be the black hole?" I asked. "What do you mean by that?"

"Well, isn't a black hole a singularity? How can it be shaped like a bagel?" The black hole theory that I was familiar with, at the time, described a black hole as a mathematical "singularity"; that is, an object that was so gravitationally compressed that it actually ceased being a three-dimensional physical object. I seemed to recall that it was theorized that the black hole could be a torus (donut—or bagel—shaped), but in only two dimensions, with zero thickness. I could never really fully understand how something like that could happen, but I bought into it, I suppose, nevertheless. Black holes were weird, like relativity itself, so after a while you just don't question things too much.

Merle smiled at my "singularity" question. "Well, Ken, mathematics can lead us into some strange assumptions about any physical process, if we don't fully account for all that is going on."

"I guess so." I quickly realized that Merle was probably about to tell me that the whole idea of a "singularity", in regards to a black hole, was actually incorrect.

"The whole idea of a black hole being a singularity," Merle said, "is actually incorrect. The basic nature of the black hole is being fundamentally misconstrued. For example, would you say that the black hole is gaseous, liquid, or solid, or what?"

"Well, that is pretty obvious. It has to be solid. It's the densest thing in the universe- denser than a neutron star, even." At that point in my life, I really hadn't gotten into cosmology or astrophysics too deeply, and I really didn't know very much about neutron stars. The fact that neutron stars were not solid was already known, to astrophysicists. "The part about it being denser than a neutron star is obviously true. But the part about it 'obviously' being a solid is not at all true. In fact, even a neutron star isn't a solid."

"How so, Merle?"

"Well, let's look at our pancake and bagel, here. The pancake—the accretion disc— is rotating, as the material is spiraling into the black hole, represented by this bagel." Merle rotated the pancake and bagel together, as a unit.

"OK."

"Mass-energy that enters the primary body of the accretion disk begins by slowly rotating around a central point."

"OK."

"One key difference between an actual accretion disk and this pancake is that, in an accretion disk, mass-energy spirals in towards the center. As mass-energy continues to revolve around the central black hole, it gradually spirals in closer and closer, and accelerates while doing so. Closer to the black hole, mass-energy takes less and less time to complete a revolution—much in the same way that water running down a drain accelerates as it goes down, in a funnel shape.

The accretion disk actually behaves like a fluid, in a macrocosmic sense-- unlike this pancake."

"So where does the accretion disk end, and the black hole begin?"

"That's a good question. One way to think about it is that the black hole begins on the other side of the event horizon, where the incoming mass-energy has been gravitationally squeezed into a superconductive superfluid, at the greatest density possible. Due to the spinning motion of the black hole, a massive magnetic field is generated along its central axis, extending outward from the black hole, in both directions. A separate magnetic field is also generated on each side of the accretion disk, perpendicular to the central axis, in the shape of a jumbled torus (donut shape). Obviously, these magnetic fields are almost unfathomably huge. So basically, it is one vertical magnetic field that passes through the center of the black hole, and through the center of the two torus-shaped magnetic fields. The black hole is in between the two torus-shaped fields.

"These tremendous magnetic fields produce a powerful effect on our black hole, here." Merle pointed to the bagel. "They transform our apparent singularity into a three dimensional torus shape—the bagel shape, with its small hole in the center, where the vertical magnetic field lines pass. The magnetic fields are what keep the superconductive, superfluid plasma from collapsing all the way down into the theorized 'singularity'. On Akeethera, we refer to the establishment of the black hole's magnetic field lines, and the fully formed torus-shaped black hole that results from that, as the 'raising of the tent poles.'"

"So how does new matter incorporate itself into the black hole?"

"That's another good question. Basically, it's an edge-on process. As mass-energy in the accretion disk approaches the black hole, it eventually compresses to its maximum and moves across the event horizon area and into the outer perimeter of the black hole itself, beginning an eons-long spiraling journey towards the center of the black hole. As this new mass-energy enters the black hole, a roughly equivalent amount of ancient plasma that has already made the full journey to the center of the black hole gets ejected out of the central opening, and back out into space, via the powerful outgoing central axis magnetic field." To illustrate, Merle drew his finger over the top of the bagel in a spiral, and when it hit the central hole of the bagel, or black hole, he drew his finger back up in the air, rapidly, almost striking the lamp which was suspended over our table on a cable.

"The nice thing is, once you enter the black hole, there is no real passage of time due to the outrageous gravitational forces. So in the blink of an eye, you're at the inner wall of the central opening, and heading back out of the black hole. To an outside observer, though, it actually took untold eons of time for you to make that journey. The problem is that nobody could survive the ride through the black hole.

"Now, I should mention that things can get a little messy, around the top of that black hole. A lot of times massive gas clouds will be created, which can drift down and sneak directly into the accretion disk, from above, and below, which can degrade the magnetic field lines, somewhat. If enough of that happens, it could degrade the magnetic field lines enough to muck up the whole works. Otherwise, though, any material drawn into the black hole eventually leaves the black hole, via the central axis. As mass-energy finally reaches the innermost wall of the black hole, at the central opening, after eons of spiraling inward, it is suddenly and quite violently ejected into space by the magnetic field lines. Those are the relativistic jets of plasma that have been observed by your astronomers. The intertwined magnetic field lines flow out of both ends of the black hole, through the hole, extending far out into space on either side, like thick ropes."

"OK. I can actually picture that," I said.

"That is what's really happening, inside of a black hole. Remember this, later, when we discuss the hyper-dimensional nature of our galaxy, and the hyper-dimensional large-scale structure of the universe. That's when it really gets interesting!"

At that point, Lillian came by again, to check on us, and she peered down at Merle's bagel and pancake sandwich. "Is there something wrong with your bagel? Or pancakes?" she asked, again with an expression of exaggerated confusion. "Was the hole too big, after all?"

"No, not at all!" Merle told her, quite jovially. "It's perfect, absolutely perfect!"

"But you haven't even touched it," she said.

"Oh, I've touched it, all right," said Merle. "I just haven't eaten any! But everything is just fine! Thank you very much, again!"

Lillian looked even more perplexed, and then quizzically looked over to me. I'm not surprised that Lillian was one of the few people who never doubted that I was telling the truth, after I publicly revealed that Merle was, in fact, an alien. "And how is your food, Ken?" she asked. Her facial expression, as she asked about my food, was memorably questioning and uncomfortable, and her voice sort of fell off into a whisper by the time she said the final word, "Ken."

"Oh, it's very good, Lillian, very good. Thank you." I smiled, weakly, and looked away. It was a completely failed attempt at being casually reassuring that nothing odd was happening.

"OK. Well, let me know if you need anything." She returned my unconvincing smile with one of her own.

"We certainly will!" said Merle, with total, friendly sincerity.

I was relieved when Lillian walked away, I have to admit. "Merle, why can't we just get into the large-scale structure right now? We have time."

Merle smiled at me. "Well, of course we have time. I suppose you're right, Ken. All right then, you are ready to keep going forward?"

"Yes, yes, yes, Merle. Yes."

"I guess that's a 'yes', then."

"Yes."

# CHAPTER
# 27

Merle was enjoying my enthusiasm with the subject matter. "OK, well, we've already established that there is no such thing as a 'speed limit' in the universe."

"Right."

"The material on the innermost portion of the accretion disk has accelerated to a significant percentage of the ratio of space to time, relative to an observer on an outer arm of the galaxy, as it enters the black hole region itself. When we, here on Earth, peer towards the center of our galaxy, the mass/energy there is moving at a significant percentage of the ratio of space to time."

"OK."

"The material continues to accelerate, to nearly the ratio of space to time, as it spirals faster and faster towards the center of the black hole, before it is ejected, top and bottom, via the magnetic field lines."

"OK."

"Now let's think about a highly advanced civilization in the galaxy, which has the ability to pilot a ship in towards the black hole, spiraling in the same direction as the accretion disk, while accelerating even faster than the surrounding material in the accretion disk."

"OK."

"They are able to accelerate so powerfully, in fact, that they are able to accelerate towards the ratio of space to time, relative to an observer on an outer arm of the galaxy, before they get too close to the black hole. At that velocity, their centrifugal force helps to keep them from falling into the gravitational trap of the center of the galaxy."

"All right."

"They just have to avoid colliding with other material in the accretion disk."

"Of course."

"At that velocity, they will begin to have difficulty seeing an observer on an outer limb of the galaxy, due to their velocity difference, and the effect of length contraction, which we know from the Lorentz transformation equation."

"Because they are approaching $c$, relative to the observer. That sounds right."

"So let's kick it up a notch, and think about what happens if the ship continues to accelerate, beyond the ratio of space to time, relative to that observer, as you now know is entirely possible."

"Sure."

"Will the ship be able to see the stars, and the observer, on the outer limb of the galaxy?"

"No. Not if they are traveling beyond light speed... excuse me-- beyond the ratio of space to time, relative to the observer."

"True. But will they be able to see the accretion disk?"

"Well, I guess they could still see at least some of the innermost portion of the accretion disk, which is probably still within their space/time reference frame."

"True! Very good, Ken!" He paused and looked around for a moment, and then leaned across the table, closer to me, and whispered. "So are you ready for the big reveal about the galaxy?"

"Yes, Merle, yes!" I think he definitely enjoyed teasing me just a bit, at times.

"Well, as the ship continues to circle the black hole, accelerating beyond the ratio of space to time, relative to the original stars on the outer limb of the galaxy, the crew members begin to see some interesting things." He smiled at me, rather enigmatically.

"What? What do they see?"

"Well, when they look inwards, towards the galactic center, they can still see what you would call the 'escape horizon' area of the black hole, looking very much like it always did, albeit distorted, in a relativistic sense. And they can see the contracted remnants of the accretion disk, which they now greatly exceed in velocity. But when they look outwards, away from the center, they see something amazing, which is also greatly contracted."

"And what is the amazing thing they see, Merle?"

"First, looking outwards from the center of the galaxy, they are surprised to see a very large gulf, perhaps millions of parsecs across, where there were basically no stars whatsoever. Beyond that, however, there were stars again, well outside and beyond where an accretion disk would be. These are different stars, flowing outwards in spiral arms, in the same manner as the previous stars, where the ship originated from. And these new stars are traveling much *faster* than their ship is traveling, in their common direction of travel, so the stars are visually contracted, as we know from the length contraction transformation."

I sat there and thought about it for a moment, a bit confused. "How can the stars be different? And how are they traveling faster than the ship now?"

"Because this galaxy-- and every other galaxy in the universe, Ken-- is hyper-dimensional. And these new stars are rotating around the central black hole at velocities that are about twice the ratio of space to time, in comparison with the original stars, where the original observer resided."

"Oh." The thought of our galaxy being hyper-dimensional seeped in on me, gradually. "So these literally are different stars, traveling at a different velocity than 'our' stars. So why can't we see them?"

"Because of the velocity difference."

"Oh, right. I knew that. Well, what happens as the ship continues to accelerate?"

"Why don't you tell me, Ken?"

Sometimes I hated it when Merle compelled me to think. But I sat there a bit, and took another sip of coffee, while I thought about it.

"Well, as the ship continues to accelerate, eventually they will match the rotational velocity of these new stars, so they will share the same reference frame, and the stars will no longer visually appear to be contracted."

"Exactly! And what about the old accretion disk?"

"Well, by that point, they wouldn't be able to see much of it anymore, if at all, since they'd be traveling at a velocity greater than the ratio of space to time, relative to the accretion disk."

"True. But what about the new stars?"

I didn't know what Merle was trying to say. "What do you mean, Merle?"

"Well, don't the new stars also have an accretion disk?"

I thought some more. "I suppose so. I suppose they would have to."

"And how fast is this accretion disk rotating, compared to the original accretion disk?"

I took another sip of coffee. "I suppose, if the new stars are traveling at twice the ratio of space to time, compared to the old stars, then the new accretion disk must be rotating at twice the ratio of space to time, compared with the old accretion disk. So if the ship has accelerated to match the velocity of the 'new' stars, then observers on the ship would then be able to see the 'new' accretion disk, as well. But nobody on one accretion disk could see anybody on the next faster or slower disk. In fact, nobody *anywhere* in one dimension of the galaxy can see *anybody* in the next dimension over, because of the velocity difference of $2c$. Not even close." I know I broke out into a grin as I pictured how "our" accretion disk sits, like a giant plate upon the fabric of space/time, within a stack of other accretion disks. Each disk, traveling in its own continuum of velocities, has a complete physical dimension of the Milky Way galaxy surrounding it, thereby insulating it from any direct contact with the inter-dimensional disks. There are important indirect effects, though, as it turns out.

Merle, meanwhile, was grinning from ear to ear, almost literally, and he gulped down some tomato juice in his excitement. "Exactly, Ken. Exactly. That is completely correct."

"But yet both disks are connected, via the multi-dimensional black hole."

"Exactly, Ken! Exactly! The black hole is the bridge to all of the many, many accretion disks that comprise the galaxy. Really, what we are talking about is a complete hyper-dimensional set of accretion disks."

"Wow. I can really see that."

Merle leaned across the table again, and looked me squarely in the eyes. "So then what happens?"

"What do you mean?"

"Well, what happens if the ship begins to accelerate, back in the direction of the new accretion disk?"

I thought about that some more, and finished off the last bite of my eggs. As I swallowed, it occurred to me. "The same thing would happen, I suppose. If the ship continues to accelerate, eventually the disk would fade, and a new set of stars would come into view. Then the new accretion disk comes into view. The process repeats."

"Yes, yes, yes, Ken, yes! And how many times could that process repeat?" The only logical answer, bizarrely, seemed to be "forever", so that's what I said.

"Forever, Merle?"

Merle exhaled, smiled at me, and slumped back into his chair. He reached out and took another sip of juice, which he did seem to enjoy. "Actually, no. We don't believe that galaxies extend throughout infinite time. We believe that, after an enormous number of dimensional bands of accretion disks, a galaxy reaches its limit, hyper- dimensionally speaking. Still, it's quite certain that the multi-dimensional galaxy involves a vast dimensional range of linked accretion disks—probably millions, or billions, or perhaps trillions, for a massive galaxy-- all circling around a multi- dimensional, central black hole."

"Wow. So less massive galaxies exist through a smaller dimensional range?" The thought of this new information was quite mind-blowing to me.

"Yes, that is what seems to be the case. But regardless as to the number of dimensions the Milky Way truly inhabits, rest assured that the portion of the galaxy that we can see is but a tiny portion of the multi-dimensional totality. There are certainly a very large number of bands, all circling the same black hole, and all separated from each other by a velocity of about twice the ratio of space to time in the universe."

"Why twice the ratio?"

"Any slower than that, and it wouldn't work. Any faster than that, and it wouldn't work. And you might have noticed that the universe tends to have a way of working out. So $2c$ it is."

"Wow, Merle." Suddenly, our galaxy had gotten billions or trillions of times larger, and the comparison of this newly realized behemoth with our own tiny Earth was creeping in on my consciousness like an elephant in an elevator. Contemplating the enormity of the hyper-dimensional Milky Way was actually making me feel a little bit woozy.

"The black hole is the great leveler of the situation," Merle continued. "The superfluid black hole exists at the entire full range of velocities of the greater accretion disk, and all the velocities in between. The hyper-dimensional black hole connects the entire thing."

Merle dabbed his mouth with his napkin, and smiled at me. "Now, after our ship moves forward into another dimension of space/time, how do we get it back? How do we travel twice the ratio of space to time *slower*?"

That one really made me think. I finished my last bite of toast. I took a few sips of coffee, but it wasn't coming to me. "I have no idea, Merle. I'm not sure how you could reverse it. Is that even possible?"

"Of course it is! The universe usually makes things easy, Ken. One way to do it would be if our friends in the ship reverse course, quite literally, and travel back against the rotation of the disk, faster and faster. That would eventually take them back to their original dimension of space/time. You see, Ken, in the universe the concept of 'faster' or 'slower' isn't always valid. You never really travel 'faster' or 'slower', you just travel in different reference frames of space/time, in either direction."

Truthfully, I really didn't quite get it at that point. It seemed very strange not to think of velocity as being 'faster' or 'slower'. Again, though, after more thought and discussion, it began to sink in. Now, it seems strikingly obvious to me. Velocity is meaningless unless it is relative to something else measurable. In relativity, velocity is just a way to establish a space/time frame, relative to the space/time frame of something or someone else.

Merle sat back and put his napkin back down on the table. "This was a nice little discussion, Ken. I really enjoyed it. But I'm afraid we will have to talk about the hyper- dimensional mega-scale structure of the universe another time. The girls are waiting for us in the parking lot, and your friend Kim will be getting off work soon, too."

"That's right! I almost forgot about L.C.!" I said.

Just then, Lillian came by with a container for Merle's left-overs, which I packed up after I separated the bagel from the pancakes and added some butter and syrup to the pancakes, and cream cheese to the bagel. I ate them eventually, later that night. Merle wouldn't let me pay, so he paid the bill and left a nice tip for Lillian, and we headed out to the parking lot. I had no idea where he was getting the money from, but I saw he had a well-stocked wallet—cash, with quite a few 20s and tens in the mix. Later, Merle told me that he actually received the cash from sympathetic alien sources. These sources, according to Merle, actually worked Earth jobs, and received paychecks, as part of their on-planet mission. That little fact really freaked me out, at the time, I have to admit.

# CHAPTER 28

Out in the parking lot, a snazzy little red two-door sports car was parked to the left of Merle's nondescript vehicle. I could see Atropha behind the wheel, and Clotro in the front passenger seat. As we approached, Clotro opened the passenger door, leaned forward, and unhinged her seat, so that Latsis could clamber out of the back. Latsis reached back into the car for something fairly large, which had been on the seat next to her. She pulled out a blue cat carrier- a cage-like enclosure that is used to transport cats.

"Sorry to interrupt-- we knew you boys were busy in there," Latsis said, smiling at us. "I've never seen a bagel and pancake galaxy before, Merle. That was cute. But you're going to need a cat carrier, Ken, to take L.C. to the vet." She held the carrier out to me, and I accepted it.

"Thanks," I said. "I didn't even think about that."

"I know," she said. "That's OK. Remember, I told you that I was going to help you. Here, I have something more." She reached back into the car, and pulled out a cat litter box, a bag of cat litter, a scoop, several cans of cat food, and three cat toys. "L.C. likes canned food, also. And I'm sure she'd like to play with some toys."

"Wow. Thanks!" I said. I instantly realized that I probably should have been a lot more proactive about the preparations, and I was a little embarrassed with myself. "I didn't realize this was going to be so complicated."

Latsis sort of gave me the eye a little bit.

"Not to say that it's complicated," I corrected myself. "This is going to work out well, I think."

"I know!" Latsis said. "I really do believe it will."

"Yes," I said, "I do believe it will." I was going to have a little pet-- at least temporarily—and in spite of my concerns, it did sound somewhat appealing, if the cat was friendly. I might even develop a nice friendship, on some level at least, with Kim. That sounded even more appealing. So I actually was feeling a little bit hopeful, in spite of myself. The downside, of course, was what seemed to be the much more likely outcome- a double rejection, from both the cat and from Kim. Normally, that fear would have easily kept me from even making the attempt.

I saw Atropha reach her arm out of the open driver's side window, and give Latsis a wave of the hand. That was the signal. "OK, gotta go!" said Latsis. "Good luck, boys!" And with that, Latsis slipped back into the car. Clotro swung the door shut while Atropha started the engine, and they pulled out of the space and headed out of the parking lot. In a few short moments, they were gone.

"We'd better hustle, also," said Merle. "Kim is going to be there soon."

When we got to the house, we unloaded the car and brought everything into the house, except for the cat carrier. "Leave that out here with us, on the porch," Merle said. "We're going to need it."

"Now what?" I asked.

"We wait for Kim," said Merle. "She's almost here."

I was a little restless, waiting, and Merle seemed somewhat amused, as I paced back and forth. "Ken, are you nervous?"

Truthfully, I actually was a little nervous. But I didn't really want to admit that to Merle. "It's just that I've never had a cat," I said. "I know a little bit about dogs, but I really don't know what to do with a cat. I have no idea how we're going to get L.C. into this cat carrier."

Merle laughed. "Well, I promised Latsis that I would help you, so don't worry. First of all, it helps to know that cats are somewhat different than dogs."

"I figured."

"Cats can be a little more sensitive, in some ways. Avoid any fast movements, at first, to gain her trust."

"So how do we get her up here?"

"That's easy. We have food, and she must still be quite hungry. Once we open a can of cat food up here, and call to her, she should come running, basically."

We talked some more about cats, and Merle gave me some ideas which were actually quite intuitive and helpful. He even pulled a small device out of his pocket, which projected some holographic videos—instructional-type videos, really, of people interacting with cats, so I had some idea of how to pick them up, and hold them, and that sort of thing.

"I've seen a lot of cat videos," I said, "but not these kinds of videos."

Merle laughed at my joking reference. "These are much more helpful than watching some cute kitten playing with a ball of yarn," he pointed out.

"True." I was super interested in the holographic projector, but before I had the opportunity to ask Merle if I could check it out more closely, Kim drove up and parked her car at the curbside. Merle quickly put the device away, to my great disappointment. I was also going to ask him if we could watch one of those Liszt videos that he had mentioned before.

Kim came walking up to the porch. "Oh, wow!" she said. "You have a cat carrier! I didn't even think of that!"

"Ken's present to your Grandma Neddie," Merle told Kim.

"Thank you so much! You didn't have to do that, though!" Kim protested. "Let me pay for it, at least."

"No, please," I said. "That won't be necessary. And don't thank us until we actually manage to get L.C. in there." I was embarrassed that I didn't actually buy the carrier, or even think about it, for that matter, and I was still doubtful that Merle's scheme would work, anyway.

"How *are* we going to get her in there?" asked Kim.

"We're going to open a can of cat food, and hope she comes for it," I told her. "Do you think it will be that easy?"

As it turned out, and to my amazement, it was fairly easy. Merle put the can on the top rail of the porch. He opened the can, as loudly as possible, and spooned it into the food dish, clanking against the ceramic dish with the metal spoon, again as loudly as possible. Merle then suggested to Kim that she call for L.C., which she did. In less than a minute, L.C. hopped up onto the porch, and went straight to the food dish to eat.

Merle suggested that I slowly move closer to L.C. and try to pet her as she ate. She was hungry enough to let me pet her. "Let her finish eating, while you pet her occasionally," said Merle. "As soon as she finishes, pick her up and place her into the cat carrier." Merle walked over to the carrier and opened the door.

He made it sound so easy, but when he suggested that I pick her up to place her into the carrier, I almost panicked, even though the instructional hologram showed me how to pick up a cat. I imagined L.C. viciously slashing me, as I made the attempt. Luckily, I didn't have much time to fester on my fear. L.C. finished a moment later, and stepped back from the dish. It was now or never, and I picked her up and scooted her into the carrier. She went in nice and easy, with maybe just the slightest squirm, and I shut the door behind her and closed the latch. I shuddered a bit, involuntarily. That was the first time I had ever petted a cat, let alone picked one up, and I was very glad it went so well.

"Yes!" shouted Kim. This time she didn't hold back, and she bounced over to give me a very big, very sincere hug. "Thank you so much!" she said. "I can't believe you just did that!"

"I can't, either," I said. Although what I probably meant was that I couldn't believe she was hugging me, and so emphatically, at that. I hugged her right back, and even lifted her off the ground a bit, in the process.

We both stepped back, maybe just a little embarrassed over the enthusiastic hug over the simple act of getting a cat into a box, and we both laughed at ourselves. "Sorry," she said. "I got a little carried away."

"No problem," I said.

"I think he liked it," said Merle, with a wink.

Kim and I both laughed at that, and we both bent down together at the same time to see how L.C. was doing in her carrier. We bumped our heads together, in the process, and had another laugh.

"We are a regular comedy routine!" Kim said.

I suddenly flashed back to Latsis' words of encouragement, and I blushed a bit, at how well things were going, so far.

"Look at her!" said Kim. "She doesn't look very scared!"

It was true. L.C. was already sitting down, in the carrier, and looking out the front door. She began to groom herself, apparently satisfied after a nice meal.

"Looks like a natural," said Merle.

After that, everything seems almost like a dream, as I think back on it. Merle took his leave of us, and I went to the veterinarian with Kim. I held L.C. in the cat carrier on my lap, as Kim drove. We checked in at the front desk and had a ten or twelve minute wait, before L.C. was called. Kim and I had a nice conversation. We actually had a lot in common, especially since we were going to the same school. Also, we lived fairly close to each other, on campus, as it turned out. It was amazing that while we had never met on campus, here we were, back in Rundle Heights, at the veterinarian together, just hours after first meeting each other. To me, it felt like a very unusual first date, and an exceptionally good one, at that. So far, Kim had given me no indications that she was in any sort of relationship.

After L.C.'s checkup, and after receiving a recommended vaccination or two, the vet gave us a synopsis of her situation. L.C. appeared to be perhaps one year old, and her wound was not very bad. As Merle had mentioned, it would heal quite easily. L.C. was, indeed, a female, which did not surprise me. But we were surprised to find out that L.C. had been spayed, and she also had been declawed in her front paws.

"That's probably how she hurt herself," said the vet. "She's not going to be able to climb fences or trees very well without her front claws, and she probably took a fall onto a fence or a branch. She won't be able to defend herself, either, without the front claws. This cat should definitely not be an outdoor cat."

The vet told us that, although L.C. obviously must have been somebody's pet, she wasn't micro-chipped, and didn't have a collar, either, so there was no way to identify her owner. Nobody had called the animal hospital about a missing cat that fit the description, either. He suggested that we post some signs in the neighborhood explaining that we found a missing cat. L.C. had some easily identifiable markings, so if it was somebody's cat, they should be able to describe her quite well. On the other hand, he mentioned, L.C. most likely was in need of a good home, and it was quite possible that her original owner wasn't interested in getting her back.

We got L.C. back home, and set up her litter box, first thing. It only took her about two minutes to discover it and use it. I filled a dish with some dry cat food and placed it in the corner, and L.C. again didn't waste any time getting into it. I offered to make some tea, and by the time I came out to the living room with the tea, Kim was petting L.C., who was arching her back and flouncing around Kim's legs. "She is the sweetest little thing!" said Kim.

Kim mentioned that she should call her grandma Neddie, to tell her the good news, and to ask her if she'd like us to put the "Found Cat" signs up. I don't think either of us was in too great a hurry to put the signs up, at that point, but Kim called her grandma in the hospital, and Neddie agreed that the signs should go up. "She is so happy!" said Kim, after she hung up with her grandma. "Although I think she hopes nobody responds to the signs."

Kim and I sat around and finished our tea, while we got to know L.C. a little better.

L.C. wasn't settled in enough to play with her cat toys yet, but she watched Kim and me bat them around, trying to entice her. "We look ridiculous!" Kim observed, and we both laughed loudly when L.C. let loose with a loud "meow", as if she agreed we looked like a couple of fools.

After a while, Kim mentioned that she had to get back home. I immediately guessed it was a boyfriend thing, but she surprised me when she revealed the reason. "My mom is waiting for me to go with her to the hardware store, to help pick out paint colors for my room. My dad wants to start painting next weekend." Well, that was a relief! So we quickly made a few "Cat Found" signs, and we went out and drove around the neighborhood to hang them up, before Kim left. As we posted the signs, I was feeling quite hopeful about my budding friendship with Kim, while at the same time I found myself actually hoping, sincerely, that nobody would claim L.C. back from us.

"I hope nobody claims L.C.," said Kim. "You were reading my mind," I admitted.

"My grandma would be so disappointed. Happy, you know, for L.C., and her owners, but—you know."

"Yes. I know what you mean." I was just surprised at how quickly I was becoming attached to a cat, of all things.

# CHAPTER 29

After Kim left, I found that I actually had some time to myself. I didn't have to work that night, and I had no idea what Merle was doing. So L.C. and I just hung around the house. I made myself some food, and spooned a little more canned food into L.C.'s wet food dish, which she ate, before she ate some more dry food from the other bowl. She was skinny, but she was making up for lost meals. I did some cat litter maintenance, also. I discovered that L.C. loved to play with her furry little toy mouse that rattled when you shook it, and I howled with laughter as she scooted around the living room, chasing the mouse toy with full-out, tumbling exuberance.

Eventually, L.C. crashed out on the couch next to me, no doubt exhausted from the trip to the vet, and the food, and the play time, and I moved over to the table and caught up with my writing. I was amazed at how much had happened to me, and how much I had learned, in such a short amount of time! I got to sleep early that night, and in the morning I woke up to L.C. sleeping on the bed next to me. I petted her, while she purred, and I mulled over my plans for the day. Suddenly it occurred to me that I had agreed to go to church with my mom and dad in a couple of hours, so I rolled out of bed to start my day.

After feeding L.C. and myself, making coffee and taking a shower and dressing, I thought about Merle, and even looked out on the deck for him. I was nervous, not knowing where he was, but before long my parents came by. They were both quite surprised that L.C. was already settled into the house. I was glad that the cat didn't seem to mind having company, because I had read that some cats will hide themselves away if people come over. Neither of my parents had any experience with cats, so they asked a lot of questions, and they were greatly entertained when I showed them how L.C. plays with her mouse. Before long, it was time to leave for church.

After church, when my parents dropped me back at the house, my dad stepped out of the car and looked around a bit. He mentioned that the grass was getting long.

"I know," I reassured him. "I'm going to cut it, after you leave. And I'm going to cut Neddie's lawn, also."

My dad looked over at Neddie's lawn, which also was getting long. "Doesn't she hire people to cut her lawn?"

"Actually, no. She still cuts it herself. I've seen her."

"Really? Well, good for her. And good for you, also, Ken. I wonder when she's getting out of the hospital."

"Maybe sometime this week, I guess, depending on how things go. I'll probably keep cutting her lawn for a while, because even after she comes home, I think she may have to take it easy with that foot. It'll probably be in a cast or something."

"True, I imagine it will be. O.K., then, son." My dad gave me a little hug around the shoulder, which was sort of unusual for him. "We'll see you later." As he headed back to the car, he stopped and turned back to me. "Are you going to be talking to Kim again?"

"I guess so. I suppose she may want to check on L.C."

"That's good. All right, see you later, then."

"Bye, Dad."

"Bye. Love you, Ken."

"Love you too, Dad."

After my parents left, I went to the garage to get the lawnmower. I filled it with gas, and first went over to Neddie's house. I cut her back yard first, and was half-way through with the front yard, when I noticed Kim pull up in her car. I turned off the lawnmower as Kim came walking up.

"Ken! What are you doing?"

"I'm cutting the grass."

"Why?"

"It's getting long."

"But my dad was going to come over later to cut it!"

"That's OK. You can tell him I took care of it already."

"Wow." She looked pretty surprised, in a good way, and she peeked around the side of the house. "And you already did the back yard?"

"Yep. It hasn't rained for a while, either, so I'm also watering her vegetable garden back there while I cut the front yard. When I'm done here, I'll move the hose and water the flowers in the front, while I cut my cousin's lawn."

"Wow. That's great! You are awesome! I'll tell my dad. Thanks, Ken!"

"You're welcome."

"How is L.C.?"

"She's amazing! You should see her play with her toy mouse! She's very cute. And when I woke up this morning, she was sleeping next to me in the bed."

"Wow! That's fantastic!" Kim looked at me, mischievously. "Are you sure you'll be willing to give her up to my grandma when she's ready?"

"It won't be easy."

"I'll bet! O.K., I can see you're pretty busy now. I'll let you get back to the grass cutting. Are you going to be around later? I'll give you a call."

"Sure, give me a call."

"O.K., I will."

"Great!"

"Well, bye, Ken. We'll talk later, then."

"O.K., Kim. That sounds great. Talk to you then!"

As Kim drove off, I noticed somebody walking down the sidewalk on the other side of the street. She was wearing a hoodie, but I think it was the walking stick that gave her away. Latsis turned and gave me a "thumb up" with her free hand, which I returned with both thumbs. She flashed a quick smile, turned back, and kept walking down the sidewalk, without ever breaking stride. I fired up the lawnmower again, and got back to work.

# CHAPTER
# 30

After I finished cutting both lawns, and watering Neddie's flowers and then Walter's plants, I finally was able to go back in and take my second shower of the day. Then I wolfed down a quick lunch, so I would be ready in case Kim called or came over. As soon as I had finished eating and washing the dishes, I heard that deep musical-like tone again, briefly. Before I could dry my hands, the doorbell rang. It was Merle, who I had almost forgotten about, with everything else going on.

As I write this, I realize that I had never even thought to ask Merle about that tone I kept hearing. Strange as it was, I imagine that he may not have given me a straight answer, anyway. Either way, I suppose it doesn't matter, but it does seem odd to me now that I never asked Merle about it. Anyhow, I walked over and opened the door.

"Merle! I was wondering what happened to you!" I said that to him even though, as I had mentioned, I had nearly forgotten about him, as I had been quite busy.

"I've been getting some things ready. Come on, we're going to take another little trip."

I wondered what he had in store for me today. It was going to be tough to top the Himalayas, for sure. We headed outside and got into the car. As usual, Merle turned on some music. "Space Oddity", the David Bowie song, was just ending. "I was listening to this one on the way over," Merle told me. "Bowie was sort of a space alien, himself, wasn't he? I mean, even though he was an astronaut in this song, he seemed to enjoy taking on the role of an alien."

"Yeah, I guess he did." I said. "He even played an alien in a movie, I think."

"That's right!" Merle said. "He did. It was 'The Man Who Fell to Earth.'"

We continued to make small talk while Aretha Franklin belted out "Think." Then there was some kind of blues or zydeco-- Clifton Chenier, I believe Merle said it was. The singing was in French, or Cajun, or creole, I guess, so I couldn't understand the words, but it was good. Merle told me that he loved the music from Louisiana, in general, including zydeco and New Orleans style jazz, and that sort of thing. After he said that, I was sure that he was going to fly us off to New Orleans, but that wasn't the plan, as it turned out. "No, we're not going to New Orleans," Merle said. "We'd have fun there, though. Lots of great people, music, food. Wouldn't get much science done, though, I'm afraid. Besides, I have something a little more unusual in mind."

We hadn't traveled very far before Merle pulled off the street, and into the parking lot of a little strip mall. We swung around the backside of the mall and parked behind a dumpster. Merle looked at me after he parked. "Here goes!" he said.

"Where are we going?" I asked. A half-moment of darkness seemed to flash past, as I spoke.

"We're already there," he said. "We traveled a little faster today than we did Friday."

I looked out the window, and I was shocked to see that we were no longer behind the strip mall. We were parked in a small circular drive somewhere, surrounded by trees and bushes on all sides. We began to move forward, as Merle continued driving as naturally as if we had just made a left turn.

"Where in the heck are we, Merle?"

"I'm afraid I can't really tell you, this time," said Merle. "Somewhere in the United States, let's say."

I wasn't sure why he was being so mysterious. "What's the deal, Merle?"

"Today we're going to learn about the double-slit experiment," he told me. "That's about all I can really tell you."

Well, now he had my curiosity piqued; that was for sure. Merle swung the car around a medium sized institutional-type three-story building, with no identification on it, and we parked in the back parking lot. The lot was largely deserted, due to it being a Sunday, I guessed, with the exception of one other car, and a medium-sized, plain white panel truck. I couldn't tell very much from the trees alongside the road and around the parking lot, other than I hadn't noticed any palm trees or cactus or anything like that. For a moment I thought I saw mountains in the far distance, but then I realized it was probably just a cloud bank on the horizon. But I wasn't sure, either way. It was warm, but at that time of year we could have been almost anywhere, I suppose. As we exited the car, I noticed somebody standing near one of the back entrances. He appeared to be some sort of security guard. He looked very official-- although I couldn't really place his uniform.

I followed closely behind Merle as he strode up to the guard and handed him some paperwork. "Here's the authorization," Merle announced.

The guard glanced at the paperwork for a moment, and then extended a hand outwards towards Merle. "It's nice to see you again, Mr. Akeetheran. The lab is ready for you."

"And is that the truck?"

"Yes."

"Tell them to bring the package in, then."

The guard motioned towards the truck, and two serious-looking men got out from each side of the cab. They also had uniforms, of a different sort, but still rather official- looking. I watched them go around to the back of the truck, open the door, and pull a very long and cumbersome-looking package out. It must have been over 10 feet long, maybe two feet in width, and maybe two or three inches thick. The long box seemed like it wanted to flex, and the delivery men struggled just a bit, to keep it from bending. Merle gave them a "follow us" wave of the hand, and they followed us into the building, carrying their long package, as the guard held the door open. I was completely in the dark as to what was happening, but it didn't seem like a good time to ask questions, so I didn't.

"We'll have to take the stairs," Merle said. "We'll never get that package into the elevator." And so Merle took us down a long hallway. We eventually turned left, and found a large, institutional-style granite stairway, leading upwards, from that point. "It's on the third floor," said Merle. "Please try not to bend that package very much, if you can help it."

"I'm not sure we can get it up the stairs without bending it," said one of the men from the truck.

"Oh, don't worry," said Merle. "I know we can." He was right, of course. It was a struggle, but the men were able to wind their way upwards while spiraling the package along, up to the third floor. I could tell that Merle was noticeably nervous about them possibly bending it too much, or dropping it back down the stairway.

We walked down another long hallway on the third floor, and we came to an unmarked door. "That's good," Merle told the truckers. "You can set it down right in the hallway."

"Won't you need help bringing it into the room?" asked one of the men from the truck. "It's pretty heavy."

"No, no, we're good. We're only taking it a short distance," Merle assured them. "Thank you very much for your help." I was surprised to see Merle hand each of the men an envelope, which looked like the sort of envelope that just might contain some cash.

"Thank you, sir!" both men said, crisply, and nearly in unison. They were clearly quite pleased with the envelopes, even with contents still unseen. I half expected them to salute, or maybe offer some parting comments or something, but, instead, they simply exchanged knowing smiles with Merle.

"You are very welcome," said Merle. "We very much appreciate the help."

And with that, the two delivery men turned and were on their way, quite directly. The room was some type of laboratory, and a fairly large one-- I could see that immediately. We brought the box in, and laid it down in the central area. I pulled out my pocket knife. "Should I open it, Merle?" I asked, referring to the box.

"No, no," said Merle, a little bit excitedly, as he waved his hands in the air for emphasis. "I mean, I'll open it. I don't want to make much of a mess, if we can avoid it." He pulled a little tool out of a pocket. It was almost the same size as my pocket knife; maybe just a slight bit larger. Merle touched a button or two on the device, and then he crouched down at one end of the long box. He touched the tool to a corner, apparently without applying any significant pressure, and quickly ran it along one edge of the carton. Then he quickly traced it along the parallel edge, as well as tracing a line between the two edges, where the two flaps of the corrugated box met. The now opened box revealed the incredibly clean and neat cutting of Merle's tool, which basically cut right through the packing tape without even scratching a fiber of the corrugated material.

"Holy cow, Merle! You were talking about Clotro's device the other day, but you have one that's even cooler!"

Merle, bemused and amused at the same time, looked at me. "Are you kidding me?" he asked.

"No, why?"

"I mean, this thing is just a glorified Swiss Army knife, for Gosh sakes. It's a handy little gadget, sure, but it's nothing compared to Clotro's spindle!" Merle laughed out loud. "Her device is probably a trillion times cooler!" He paused, apparently thinking about the comparison he had just made. "Heck, probably about a quadrillion times, actually!"

Merle pulled the cut flaps open more widely, and the contents of the box were revealed- thin strips of material of varying lengths, most of them the full width of the box, with various notches and slots cut into them. We proceeded to slide the contents out of the box, and onto the floor.

The material was flexible, shiny and lightweight. It had to be some kind of wonder alloy, I assumed. "What kind of alloy is this, Merle?"

Merle gave me a quizzical look. "Alloy? No, this is just styrene, I believe. 40 mil white styrene, I believe. Actually, it's probably closer to 37 or 38 mil, I'd guess, but that doesn't matter, for our purposes."

I was surprised to hear that it was just a common material. There were about 18 different styrene strips on the floor. Two of the longer ones had small appendages, or hooks, on the back.

Merle pointed over to a large piece of equipment, not far from where we had laid the styrene. I was surprised to see that it was, in fact, a large, oval pool, several feet deep, and filled with about a foot and a half of water. The entire tub, maybe 24 feet by 12 feet across, was raised about three feet off the floor, right there in the central area of the lab.

"Now here's something that's sort of neat, I guess," Merle said. He still had the tool in his hand, and he opened a little side compartment and pulled out a small blue, transparent disk, about the size of a guitar pick. He walked over to the pool, and tossed the disk towards the center of the water. I was surprised to see that the disk did not enter the water and sink, as I assumed it would. Instead, it rose upwards in a graceful arc, and assumed a spot about ten feet in the air, above what looked like the exact center of the pool. It sat there, hovering motionlessly. I thought it may have been rotating, but I wasn't sure.

I was taken aback when the disk suddenly emitted two sheets of glittery, bright blue light. One sheet plunged into the water of the pool, while another sheet glimmered down onto the pile of styrene sheets. I was startled, and I almost fell over the empty box that was on the floor, behind us. Merle grabbed me before I took a tumble, and as he set me back upright he pointed to the sheets of light. "That's our template," Merle said, and he reached down to pick up the strip of styrene which was indicated by the sheet of dazzling blue light. He handed me the strip. "Here, Ken. You take this one and clip it on the short edge of the pool, right here, where the beam is showing you."

I looked into the water, and saw that the sheet of blue light matched the length of the styrene I was holding. That's when I realized that the strips were going into the laboratory pool.

"It's OK," said Merle. "The blue light won't hurt you. It's just blue light. And the water is just plain old water." He smiled at me.

It only took about three or four minutes for Merle and I to set up the styrene strips in the correct pattern. As each piece went into place within the pool, the blue sheet of light associated with it blinked out, and a new blue sheet of light appeared, indicating where the next piece should go. I asked Merle if he had been practicing the assembly, and he didn't say "no."

When we were finished, an expanse of styrene stretched across the entire long dimension of the pool. It extended several inches above the waterline, and submerged all the way to the bottom of the pool, or thereabouts. In the center of the central piece was a single open slot, perhaps an inch and a half wide, extending all the way to the bottom, and maybe two inches above the water surface. Off of that central strip of styrene, on either side of the slot, several smaller pieces of styrene interweaved with each other in a seemingly random cross-hatched pattern, with an unimpeded, central lane about one foot wide, leading up to the single slot. Merle told me, later, that the cross-hatched sections of strips were designed to baffle and minimize any wave deflections that may have messed up the experiment, as well as to add some stability to the whole set-up.

Merle reached into his pocket and pulled out a small tin, maybe four inches by two inches, and about a half-inch in height. I assumed it was a box of mints or something, but when he opened the lid and showed me what was in the box, I couldn't have been more surprised. I bent down a bit and peered more closely at the unexpected contents. I looked at Merle and saw that he was humored by my reaction. "Merle, are those what I think they are?"

"Yes, they are!" He was laughing, now. "Miniature rubber ducks! 128 of them, to be exact! Aren't they cute?"

Indeed they seemed to be, as Merle indicated, miniature rubber ducks, perhaps a quarter of an inch in all dimensions, and perfectly cute, as he said. They were packed symmetrically into the tin, like 128 tiny yellow sardines.

I still hadn't figured out, at that point, what I was about to witness. "Merle, what is this all about, anyway?"

"Don't you see it already? We're going to run the double-slit experiment! The water in the wave pool represents the space/time continuum, and the rubber ducks represent little photons!"

"The double-slit experiment!? In a pool of water?" I asked Merle. I was suddenly very intrigued, to say the least, since Merle had told me, previously, that he was going to explain this gigantic mystery of science.

"Yes, in a pool of water, Ken. Like I said, the pool, here, represents the space/time continuum. These little ducks, here, represent little photons, or electrons, or atoms, or molecules, or whatever quantum matter you'd like. All we have to do is get the pool in motion, like the space/time continuum is. Luckily, this is a wave pool, so it's made for just that."

Again, it seemed like Merle had been practicing. He started the wave generator as if he had done it many times before, and water began moving, in a wave action, towards the slot in the wall. "Here we go", he said. Merle poured a small handful of the mini ducks into his left hand. With his right hand, Merle reached out across the water almost as far as he could reach, and he began depositing the ducks, one by one, into the flow of the wave action. The ducks bobbed on the waves towards the slot in the wall.

Some ducks missed the slot, and a couple dozen of them piled up along the styrene strip on either side, but Merle was releasing the ducks far enough out so that most ducks passed through the slot. On the other side, we had hooked one of the long styrene strips, which hung into the water and up to about six inches above waterline.

When the ducks eventually reached the opposite wall, I made a mark on that piece of styrene at the exact spot the duck hit, and then I removed that duck from the water with a set of needle nose pliers which Merle had given me. After several dozen ducks had landed, it was becoming clear that the ducks were all landing within a certain area of the wall directly across the way, as you would expect. Basically they were landing in a "fuzzy slit" interference distribution—mostly concentrated within a central portion which was a few inches wide, but with a few ducks drifting outside the normal distribution to some extent. We ended that portion of the experiment and collected the ducks back up. I noticed that, when you put the ducks back into the little tin in a jumble, they automatically self-organized and lined themselves up, symmetrically and neat, as if they were magnetic or something. That was pretty cool to see, because I didn't think they would ever all fit back in the tin.

Then we removed the styrene strip that included the slot, only to reveal the second strip behind, which had *two* slots, spaced about four inches apart. Anyone who is familiar with the double-slit experiment can probably guess what I observed when we floated ducks through this new scenario. We ran about 100 ducks through the set-up, sending them through both slits simultaneously, in bunches. Through each slit, a wave of water propagated, which, according to Merle, represented the wavelike propagation of the space/time continuum passing through the slits in the double-slit experiment. .

On the other side of the main strip with the slits, the two waves— one from each slit-- intersected and interfered with each other in a diffraction pattern. The action of the diffracted waves carried the ducks along. At some points, against the far wall, the wave troughs constructively interfered, and very few ducks landed. At other points, against the far wall, the wave *crests* constructively interfered, and there were quite a few ducks that showed up in those locations. The resultant pattern of marks that I made on the styrene, where each duck landed, indicated a classic interference pattern, just as you see in the regular double-slit experiment.

Then, we repeated the experiment, this time releasing only a single duck at a time, with clear time intervals in-between. Very significantly, the interference pattern still remained, since it was the medium-- in this case space/time, i.e. the water—that was interfering with itself, not the rubber ducks interfering with each other. In fact, you could visually look at the water, and you could see the complex diffraction pattern as the waves collided in their characteristic manner.

"Just like in the real double-slit experiment!" Merle told me, enthusiastically. "And do you know what happens if you start sticking a hand in the water, in one of the slots, trying to measure for rubber ducks, trying to feel for rubber ducks with your hand?"

"It messes up the results for the other slot, too?"

"That's right! The ducks pass through the other slot in the simple diffraction pattern of a fuzzy slot, since you no longer have two clean waves passing through the slots, to interfere with each other. You messed up the one wave by feeling around with your hand and ended up changing the flow of the other wave-- or the flow of the space/time continuum."

"Just like when scientists try to measure where the electrons go, in the double-slit experiment," I said.

"Right. No matter how hard they try, they inevitably disrupt the flow of the continuum in their attempt to measure the situation in that slot. Then, the second slot inevitably goes back to its singular, simple diffraction state."

"Wow." I sat back a bit, and looked again at the styrene that I had marked up in the unmistakable bands I had seen so many times in physics books, or online. It took me a minute or so to wrap my head around what I had just seen. "So the wave is the space/time continuum, not the photon," I said, repeating what Merle had told me when he did his rubber band demonstration in the Himalayas.

"Right."

I thought about that some more. "So a photon is really a particle, then. Not actually both a particle and a wave. The wave is space/time, not the photon."

"Right."

"Wow, wow." I shook my head. "So the photon pretty much rides the wave of the space/time continuum, then? When we observe a photon's 'wavelength', we're really watching it bob up and down, or ripple, as its being pulled along by the specific wave of the space/time continuum which can interact with it at the $c$ velocity, relative to the observer?"

"Right, right," said Merle. "That's a great way to think about it. It's just like a duck on the water, Ken. And every particle—every quantum accumulation of mass/energy—has its own corresponding continuum of space/time wavelengths that interacts directly with it, depending on its position, energy and other factors."

"What about people?"

"Even people have their own wavelength. It's just a very, very long wavelength that's extremely difficult to observe or measure."

"What about planets?"

"Planets, too, have wavelengths of space/time associated with them, even more difficult to measure and, again, not particularly noticeable on scales like that. Like I said, Ken, every quantum accumulation of mass/energy interacts with the space/time continuum, and the continuum interacts in the form of a wave, traveling at $c$, relative to the quantum accumulation of mass/energy. The wave is just a far more noticeable effect at smaller scales, like with sub-atomic particles, than it is with larger accumulations of mass/energy, like a bird. The math gets pretty crazy with something like a bird."

"Wow, OK. I can see that." I was remaining calm, on the whole, but I was still freaking out, inside, about the experiment. Wave/particle duality was such an ingrained thought for me that it was very difficult to concede that a photon is only a particle, and not really a wave at all. Plus, the whole thing seemed so simple to me, now, after the rubber duck demonstration, that I was feeling a little bit overwhelmed and stunned. It just seemed too easy and obvious, on some level, I guess, but then again the space/time continuum can be sort of intimidating to think about. The whole thing was sort of disorienting and disconcerting, in that respect.

"Then again," Merle continued, "the continuum also has its ways of affecting very large accumulations of mass/energy in hidden ways; our own galaxy is a good example of that."

# CHAPTER
# 31

I didn't have much time, right then, to ponder the ramifications of what I had just witnessed, or what Merle meant by his comment about very large accumulations of mass/energy, because Merle informed me that we had to pack up and get out of there. We unexpectedly had less time than Merle had originally assumed. I did take a moment to try to steer Merle into discussing other quantum particles, in addition to photons or electrons, but he was having none of it.

"We have to clear out of here, immediately. Regarding wave-particle duality, as you call it-- any particle, or accumulation of mass-energy, rides the wave. We are all but ducks on the lake, Ken." Merle reached into his pocket, I guess, and pulled out his little gadget. He clicked the gadget, and the chip that was still hovering above the pool pointed a beam into the water, leading us to the proper sequence for easy disassembly. As we pulled each piece out of the water, Merle quickly dried it with his gadget, and we re-packed the box. Merle held his gadget up in the air, and the chip floated back down and flipped itself right into the little chamber in the gadget. Then Merle used the gadget to reseal the end of the box, which I didn't think would even be possible. Merle and I took the box back out of the lab into the hallway, and Merle locked the door of the lab behind us, carefully wiping clean both doorknobs with a small cloth, as we made our way out. Then we carried the box further down the hallway, and deposited it against the wall next to the last door on the left side. Merle told me that "a friend" would take care of it. After that, we headed out of the building, and briskly headed back out to the car.

The guard was now parked in his car, over on the other side of the parking lot, away from Merle's car. The white panel truck had long since left, I assumed. We got into Merle's car and departed, after first exchanging waves with our security guard friend across the lot. As we pulled away, Merle looked over at me. "You must be hungry, Ken. Would you like to stop for a bite to eat somewhere?"

"You must be reading my mind, Merle. I am absolutely starving."

"How about a nice New Civilizations Special?" That was the name of one of the "house special" cheeseburgers at the R.H. Enterprise, with jalapenos and diced kimchee and some type of cheese. Merle must have known that I liked that one.

I was actually a little disappointed that we weren't flying off to Antarctica, or maybe to a picturesque little valley in the central highlands of Madagascar, or possibly to the pyramid complex on the Giza Plateau, or something like that. "The Enterprise again?" I asked.

"Why not?"

"OK," I said, somewhat grudgingly. "That's fine." I guess I was already getting spoiled with the conveniences of Merle's car/spaceship. "But this time maybe we can eat in the lounge." The lounge portion of The Enterprise was physically separated from the primary restaurant, with a separate entrance, a large bar, and several tables, as well. The two portions shared a kitchen area in the back, and the lounge area offered a slimmed-down version of the restaurant menu, including cheeseburgers. The lounge would also offer much less interaction between Merle and a waitress, I was thinking.

After a brief flash of darkness, I looked around for a moment, and I suddenly realized that we were back behind the strip mall, next to the dumpster, where we originally began our day's adventure. "Wow, Merle. I never even know when we're flying in this thing, when we go so fast."

Merle liked that. "Smooth, isn't it? Especially for such a little ship. I really like it.

They did great with this. It's a lot of fun. You know, it goes in the water, too."

"No kidding?"

"Yeah. I haven't had to try it, though. Probably won't get the chance."

We pulled back around the mall, and got back on the street. Before long, we were pulling up to The Enterprise.

"I think I'm starting to like this, Merle. Fly out to who-knows-where in about three seconds. Do your thing. Fly back in another three seconds. Have dinner. Nice way to travel!"

Merle liked that, too, and he flashed me his patented wide grin. "It's the only way to travel, Ken! It was considerably less than three seconds, though."

As we entered the Lounge, I could hear the jukebox churning out the traditional strains of "Baby's Got Her Blue Jeans On". I knew what that meant. "Daryl is here," I said.

"Where is he?" Merle asked.

"I don't know. I just know that he always plays this song."

Sure enough, we found Daryl and Mark, in the back of the Lounge, playing darts and enjoying a few cold ones. "You guys want to play darts?" they offered us.

"No, no thanks," I said. "We're getting some food, so we're going to grab a table." Being with Merle around other people still made me nervous. I was always afraid that either he or I would say something really stupid or inappropriate, and ruin everything. So I was sort of relieved that we had the ready-made excuse of needing a table, for eating.

Merle and I ordered our food at the bar and then grabbed one of the few remaining open tables. Actually, Merle didn't order any food—just a tomato juice—but I was quite hungry, so I ordered a New Civilizations Special with a cup of lentil soup, a Tribble Salad, and a pint of ale. I was actually relieved that Merle hadn't ordered any food this time, if you know what I'm saying.

As we sat down, I was expecting that we would discuss the double-slit experiment we had just conducted in the wave pool, probably only ten or fifteen minutes prior.

Merle had another direction in mind, though, as usual. "So what do you think is at the very center of our universe?" Merle asked me.

That question took me by surprise. I just had never really wondered about the center of the universe before, I think. I didn't really have any well-developed thought on that, so I just sort of said the first thing that came into my head. "Oh, I don't know. Wow.

The center of the universe, huh? Emptiness, I guess, after the Big Bang?"

"Nope. And nope."

"What was the second 'nope' for?"

"The Big Bang."

"What do you mean, 'nope' to the Big Bang?"

"I mean use your mind, Ken, and think about what might be at the center of our hyper-dimensional universe. And don't get too caught up on that 'Big Bang' stuff. You remember what is at the center of our own galaxy, don't you?"

"A black hole. I mean, a hyper-dimensional black hole."

"Correct. So what do you think might be at the center of the universe?"

"Remnants of the Big Bang?"

Merle slowly turned his head towards me, fixating his now-baleful eyes on me as he did so. Taking a deep breath, he issued forth a deep, melancholic sigh, apparently quite heartfelt, and looked up towards the ceiling as he slowly shook his head. He paused for a bit, apparently in deep thought-- with furrowed brows, even, I noticed-- before he suddenly continued speaking. "Ken, remember, I said not to get too caught up on the 'Big Bang' stuff. Instead, think microcosmic/ macrocosmic."

"Microcosmic/ macrocosmic?"

"Yes, yes. Remember, a hyper-dimensional black hole is at the center of our own galaxy. That's the microcosmic."

"That's the microcosmic," I repeated, and thought about it. I leaned my right elbow on the table, and supported my chin with my cupped hand, as if that might help.

Well, maybe it did help, because right then, like a slow-motion tsunami, the idea began to bore in upon me. As soon as I began to consider this new idea, I suddenly knew that it fit, so perfectly, that it must be true.

"The macrocosmic!" was all I could articulate, as the realization sank in upon me, layer upon layer. I was looking upwards, now, myself, as if I was imagining for the first time the workings at the center of the universe, and had set about trying to see it, up in the sky, even though we were indoors, and it was daytime, and the center of the universe is much, much, much too far away to see with the naked eye, let alone the largest telescope in the world, if in fact you could even say that you could "see" it, at all!

"What is the macrocosmic?" Merle asked me. He had an extraordinarily large smile on his face.

"Merle," I said. "It's a hyper-dimensional black hole. A supremely gigantic, hyper- dimensional black hole!"

"Absolutely correct, Ken!" Merle stood up, leaned across the table, and gave me a high-five. Absolutely correct!"

By then, I already saw the basic situation. A gigantic, hyper-dimensional black hole, functioning as a sort of ever-present Big Bang, resides at the center of our universe and sends off enormous jets of mass-energy in an eternal cycle. Our Milky Way galaxy, and all the galaxies that we can see from Earth, are part of that larger flow from the central black hole. It is such a massive flow that we can't easily recognize, from our perspective, that we are riding within it. However, we have already picked up on certain cosmological clues to the directional, anisotropic flow we ride within, for example the alignment of rotation axes of galaxies, polarization angles, and so forth.

Merle and I talked about the universal black hole some more, before our food got there. Merle explained to me that once mass/energy is ejected from the black hole, via the poles, it is compacted together rather tightly-- shepherded together by the gigantic magnetic field extending deep into space from each pole of the black hole. This mass/energy begins to slow down immediately, due to repeated collisions with adjoining mass-energy in the tightly packed stream.

After traveling for eons, the mass-energy has traveled far enough from the black hole so that it begins to break away from the attraction of the weakening magnetic field extending outward from the poles. It begins to curve back down, and away from the central axis, in a spiraling 360 degree radius, much like the fabric of an umbrella, except in this case with spiraling tines, arching away from its central pole. This mass-energy spreads apart and accelerates through space, as it eventually plunges back towards the central plane of the universe, drawn faster and faster, in a spiraling fashion, by the enormous gravitational force of the accretion disk and the beckoning central black hole. The broken-down magnetic field lines also are drawn back towards the accretion disk, thereby functioning as the spiraling "tines" of the umbrella. On the opposite side of the black hole is an "upside down" umbrella structure, functioning in the same way.

Inevitably, all the mass-energy that was ejected via the poles feed back into the accretion disk, and finally back into the black hole itself, before repeating the cycle. The universe itself, then, is the only true perpetual energy machine in our universe, and the black hole at the center of the universe is the only infinite black hole, most likely, in terms of an infinite number of accretion disks. For every amount of mass-energy that re-enters the universal black hole, an approximately equivalent amount of mass-energy is ejected from the inside of the hole, and back out into space, thereby maintaining an eternal equilibrium.

The Earth is somewhere in the acceleration and expansion phase of the cycle, in the umbrella "fabric" portion of the process, having broken away from the central magnetic axis, but not yet having reached the central plane of the universe. That's why, when we look around us, mass-energy is departing from us in all directions. That is how it looks when we are in the midst of an accelerating flow of material, which is also expanding through space from the original narrow jets to the cascading and expanding "umbrella" of mass-energy. All other galaxies appear to be receding from us in all directions, and the universe is so gigantic that we cannot easily discern that we ourselves are in the midst of a structured, accelerating flow.

Once mass-energy enters the central plane of our universe and begins to move back toward the accretion disk, the cumulative mass-energy will draw back closer together, even as it continues to accelerate, as it nears the narrowing throat of the "whirlpool" that is the black hole at the center. That is how we know that the Milky Way has not yet entered the central plane of the universe, or the universal accretion disk.

One of the things I struggled with, as we discussed it over a burger and tomato juice, was to realize that space is *not* expanding, as I had been always taught from Big Bang Theory. The modern Big Bang theory posited that all the stars and galaxies that clearly appear to be racing away from us-- the most distant ones at nearly the speed of light— are, in a sense, *not* actually racing away from us. According to the Inflation portion of the modern Big Bang theory (I say "modern" because it was not part of the original theory), space *itself* is expanding. So those stars that *seem* to be racing away from us are really *not* racing away at all, in the usual sense, according to Big Bang theory! In retrospect, it sounds almost silly, but I believed it myself. Most astrophysicists worth their salt believed it, also. Of course, that was before we knew that the universe is hyper-dimensional.

Eventually Merle got through to me that those galaxies we see receding away from us in all directions are, actually, speeding away from through normal physical space, exactly as they appear to be doing. "It's nothing trickier than that," Merle said. "We are all in the flow, accelerating back to a cataclysmic re-acquaintance with the universal black hole- the Mother of the Universe, you might say. Space, or space/time itself is not expanding, overall. Space/time is curved by mass-energy, and it is in a less curved state farther away from mass-energy, but all this talk of expanding space as being the sole explanation, or the cause, of retreating galaxies is incorrect.

"In fact, three-dimensional space is entirely static, throughout the infinite cosmos. It's really the curvature of the time dimension that causes the apparent "curvature" of space/time. Since space and time are interconnected, though, the curvature is also innately interconnected. It might sound a little contradictory, but it isn't.

"Gravitation, in fact, might be considered as the Fourth Law of Thermodynamics, by your parlance—the Conservation of Time. This makes a lot of sense, in light of what we know about the Lorentz transformation equations and time dilation. You know that time passes more quickly at higher altitudes, correct?"

"Yes, I do recall that."

"That's right. A person living on a mountaintop actually experiences a faster passage of time, compared with somebody living at sea level. The Earth's mass causes a displacement, and curvature of space/time. This curvature is greater the closer one gets to the Earth."

"O.K."

"So less curvature up on the mountain, where time passes more quickly."

"O.K. I see that."

"So what happens if you drop a rock off the mountain?"

"Well, gravity will make it bounce down the mountainside."

"That's right. In the universe, time is always conserved by mass/energy, as much as possible. In other words, mass/energy is pushed by the space/time continuum towards the slowest possible passage of time. The universe pushes the rock towards the slowest possible passage of time, where space/time is the most curved. That's exactly what gravitation is, so you could describe gravitation as, simply, the Conservation of Time. Remember, time is actually an energetic manifestation of the space/time continuum. Therefore it is a thermodynamic-type process, in a relativistic sense."

Merle threw so much information my way that sometimes I didn't really focus on something until well after he said it. That's how it was with his explanation of the Conservation of Time. It wasn't until sometime afterwards that I grasped how amazing it was to consider gravitation to be one of the Laws of Thermodynamics.

Regarding the central black hole of the universe, I'm not saying that it's completely inevitable to get sucked back through at the end of a cycle. Merle mentioned at some point that it was believed there are civilizations that have existed for so long that they were able to develop technologies that helped them to escape the central destruction, either by short-cutting the black hole and riding the flow back outwards, or possibly simply by going back against the flow, and avoiding ever reaching the proximity of the center. That type of wild adventure is the sort of thing I think about, sometimes, when I have some quiet time to think, maybe late at night before going to sleep. If these civilizations exist—and I believe they do—I am rooting for them to escape the clutches of the universal black hole!

# CHAPTER
## 32

Merle took another sip of tomato juice and smiled at me again. "Well, Ken, it sounds like the music has stopped."

"I guess it has."

"Well, I have a five dollar bill in my pocket."

"And?"

"And I'm going to play some music."

I sat there while Merle put the money in the juke box. As the strains of "Here Comes the Sun" poured out of the speakers, Merle came back to the table. "How about a game of pool?" he asked me. "I've never played."

"It's sort of difficult, Merle. I'm not really that good, myself. But sure, we can play."

"Good! I've watched people play a couple of times, so I have a pretty good idea of what to do."

"Good! You probably know more about it than I do, Merle."

"Oh, I doubt that," he said.

It's probably needless to say, but you might have already guessed that Merle only needed a few turns to clear his "striped" balls off the table. I think I sank only two "solid" balls.

"I thought you never played," I said, after he sunk the last four shots in a row. "I haven't! But like I said, I've watched a few games."

"Yes, I can see that." There's no question that I might have been just a wee bit annoyed at how good Merle seemed to be at sports. For him to shoot pool like that, having never played before, was almost discouraging to me.

By then, Daryl and Mark had come over to watch us play, which made it impossible for Merle and me to talk about the universe, or anything like that. After Merle and I finished, the four of us played a game of doubles- Merle and me against Daryl and Mark. Daryl was a good player, so the match ended up being mostly a duel between Merle and Daryl, with Mark and me mostly missing our shots. Daryl ended up winning the game for his team when he went on a run and sank the last three balls, finishing with a very difficult shot where he used "English" to curve the cue ball around one of our balls that appeared to be blocking Daryl's position. Rather than getting frustrated with how the game ended, Merle got excited watching Daryl shoot, and he gave him a very enthusiastic high-five after the last ball went in. "That was awesome! You made the ball curve so much by spinning it!" Merle almost shouted. "Remarkable!"

Daryl looked first at Mark, and then at me, with a look that questioned, "Is this guy for real?" I wasn't about to tell him that Merle had never played a game of pool before, in his life.

Two other guys had come in, while we were playing, and called "winner", meaning they got to play the next game against Daryl and Mark. That gave Merle and me a chance to head back over to our table, which luckily was still empty. "Three Little Birds" was playing on the jukebox as we took our seats, although at the time I didn't realize that was the title of the song. I always thought it was "Don't Worry About a Thing."

"Is this Bob Marley?" I asked.

"Yah man," Merle said, in his best Jamaican accent. "Love this song."

"It is great," I said, as we each took a sip of our respective beverages. "You're pretty great, too, at sports. I think you'd be great at any sport you want," I said.

"I always liked to play sports back on Akeethera, other than maybe swimming, which I was never that fond of. I've wanted to play a game of pool for several months, now.

That was a lot of fun. I don't know why I never had a pool table put in, on the ship. That doesn't take up as much space as a bowling lane. Maybe I'll have that done tomorrow, if possible."

"Everything is fast with you, isn't it, Merle?"

He looked at me as if he didn't understand why I was asking that question.

I just laughed. "It's all relative, I guess, isn't it?" Right then, my cell phone rang. It was Kim. I told her we were at The Enterprise, and she said that she would stop on by in a few minutes.

In the meantime, I was getting thirsty again. "I wouldn't mind having another beverage," I said.

"Well, I guess I wouldn't mind another tomato juice, then."

So I went to the bar and got us another couple of drinks. When I came back, Merle mentioned that we were getting "pretty close", now.

I was confused by that statement. "Close to what, Merle?"

"We're getting close to wrapping up our discussions of the hyper-dimensional universe. Right now you have a very strong understanding of the large-scale structure of the universe, and of Hyper Relativity. But let's think about how gravitational fields operate in a hyper-dimensional sense."

"Are you going to say that mass-energy traveling faster than the ratio of space to time—relative to us—still affects our own space/time continuum, in our own physical and perceptual space/time reference frame? In other words, we are gravitationally affected by mass-energy from outside of our own physical dimension?"

"Yes! Very good, Ken."

"Is that possible?"

"Yes, of course it is possible. Why wouldn't it be? Isn't space/time a continuum?"

"Yes."

"Well, then, if you gravitationally compress one portion of the space/time continuum, shouldn't it affect adjoining space/time as well? It's very much like pressing your thumb into a rubber ball—the spot under your thumb is not the only part of the ball that compresses down. Adjoining areas of the rubber ball are pulled down, as well. In fact, shouldn't gravity affect an ever widening shell of influence, in an ever-dwindling way, as time passes? Isn't that infinite interconnectedness the very nature of the continuum itself?"

I could picture exactly what Merle was talking about, more or less. "So this is gravitation that exists outside of the field equations?" I was referring to Einstein's field equations from General Relativity, describing gravity.

"Well, Ken, I would tell you this. You make a great observation, regarding the field equations. This new information you will be bringing to the table will, at the very least, add another level of significant depth to the understanding of the field equations, and it will, as you suspect, add more layers of mathematic possibilities, going forward. But I would suggest to you that you already have quite a lot on your plate, in terms of physics and cosmology, without delving too deeply into the field equations, yourself. There is much room to explore, mathematically, within the hyper-dimensional realm, but you need to maintain a focus. Otherwise, you will find endless additional corridors to explore along the way, and never reach your ultimate goal. The important thing to take away is that gravity bleeds over into other adjoining dimensions of the continuum, to a certain extent."

"So, which extra-dimensional mass-energy affects us the most, gravitationally?" I asked, hoping for some elucidation.

"Right here, in the Milky Way Galaxy?"

"Yes."

"I already told you."

It took me a moment. "You mean the Milky Way itself affects its own gravitational cohesion?"

"Yes, of course. Mass-energy throughout the entire hyper-dimensional range of the Milky Way compresses space/time, all the way into our own space/time reference frame. All that extra-dimensional mass-energy creates what you might call a gravitational soup. That soup is what keeps the galaxy from flying apart from its centrifugal forces."

"A gravitational soup?"

"Well, on Earth I've heard it referred to as "dark matter". *"That's* what dark matter is?"

"Absolutely. The dark matter 'halo' surrounding the Milky Way is the gravitational bleed-over from the extra-dimensional portions of our own galaxy, primarily. Space/time is truly a hyper-dimensional continuum."

Dark matter explained! I was so excited by that, I nearly knocked over my glass of ale! "And dark energy?"

"That is the flow of the universe, as we discussed. The fact that we are now in the 'acceleration and expansion phase' of the universal cycle, and accelerating back towards the gravitational attractor of the central hyper-dimensional black hole, is the simple explanation for dark energy."

"Ha!" I laughed, at Merle's description. "A simple explanation! Ha!"

"But it *is* simple!" Merle said.

"Simple for *you*, maybe, Merle." But in truth, when I thought about it, later that night after I got home, I realized that Merle was right. It was actually deceptively simple.

"I'm afraid that's all I have time for, tonight," Merle said. "I have an appointment, back on the base."

Just then, I saw that Kim was approaching the table. I had almost forgotten about her! She came up to us and sat down next to Merle. "Hey, guys! Why aren't you shooting pool?"

"We already did! We lost!"

We all said hello, and so forth, but soon Merle stood up. "I really have to get going," he said.

"Don't leave on my account!" said Kim.

I mentioned to Kim that Merle was already about to leave, before she came to the table. "Actually, I was about to head back home myself, to see how L.C. is doing," I said. "But would you like a drink or anything while we're here?"

"No, I'm good," said Kim. "I'd actually like to see how L.C. is doing, also, though, if you're going to the house."

So we said our goodbyes with Merle, and Kim drove me home. She stopped by for almost an hour. We had a nice conversation, and we played with L.C. The more we talked, the more I found myself being attracted to Kim, and I couldn't help thinking that the feeling might be mutual. She still had never mentioned anything about a boyfriend, and I think I was pretty clear about not having a girlfriend.

When she finally got around to leaving, and I walked her out to the car, I think we were very close to a goodnight kiss. I think we both wanted to, but it still seemed a little soon, and instead we just said goodnight, awkwardly. After I got back into the house, I walked over to L.C., who was sitting on a chair in the living room, relaxing after the play session. I was becoming a real pet owner, already, because I was starting to talk to that cat as if she was another person. I could tell that she liked it, too, because just talking to her in a nice tone of voice was enough to make her purr, especially if you said her name. "The next time that happens, L.C., I'm going to kiss her," I said. L.C. meowed right back at me, which I was all too happy to take as her approval of the idea.

# CHAPTER
# 33

The next morning, a Monday, I worked the breakfast shift at The Enterprise. Lillian was working that morning, also, and I tried to avoid her as best I could. I was afraid she would be asking questions about my strange friend that I wasn't prepared to answer.

Obviously, Merle had made quite an impression on her, since she tracked me down, anyway, to ask uncomfortable questions, as I had feared. I answered as vaguely as possible, and tried to deflect the questions by making a joke of them. I don't think I placated her very much. I was glad, quite frankly, when my shift ended before she could grill me further.

On my way home, I stopped by a pet supplies store to pick up some more cans of cat food, along with some cat treats and another bag of cat litter, and a few other miscellaneous cat-related items, like another pet dish and some new toys. As I walked along the street, on my way home, one of the cat toys I bought tore a hole in one of the bags, and cat toys and treats and the dish fell onto the sidewalk. I managed to cram everything from that bag into the remaining bag, but before I got very far, that bag ripped out, also, and now twice as many items spilled out. I was upset, because I couldn't figure out any way to carry all the loose items, along with the heavy bag of litter, and I was still only about half-way to Walter's house. I was just thinking about stashing the stuff in a nearby bush, or something, until I could come back to retrieve it all, when I heard the buzzing note again. Immediately, I heard a voice from the street say, "Hey, handsome!" I turned around to see Latsis hanging out of the open window of the girls' little red sports car, which had pulled over to the curb. "Need any help?" she asked.

"Oh my gosh, yes! I'm so glad you're here! My bags broke!"

"I see that," said Latsis.

"They just don't make 'em like they used to!" chimed in Clotro, from the backseat. I guess Merle wasn't the only one who enjoyed using clichés.

"Why don't we put all that in the trunk?" asked Latsis, as she got out of the car to help. "We'll drive you home."

She didn't need to ask me twice. Latsis and I gathered my loose cans and toys, and poured everything into the trunk of the car. I was extremely excited to think about what the inside of the girls' car would be like, because it seemed like they were quite a bit more advanced than Merle, even.

Upon entering the back seat of the vehicle, I got a big surprise. Their vehicle wasn't at all like Merle's. In fact, even from the inside, it looked just like a regular car. "This looks just like a regular car from the inside," I said.

Clotro laughed at that. "It looks just like a regular car, because it *is* a regular car.

We actually bought this car from a car dealership."

I found that hard to believe. "Seriously? Then how do you get it back to your ship?"

At that, Clotro reached into her pocket, I guess it was, and pulled out her amazing spindle. She held it in her hand, and looked over at me. "Oh, it's easy to get back to the ship," she said.

Well, that spindle-thing of hers was definitely amazing, like Merle said. I did notice that I never really got a good look at it again, other than the first time I had seen it. Any time I would try and look directly at the spindle, Clotro subtly changed its inclination so that I was looking at it end-on, and I couldn't really make out anything too clearly from that angle other than a very shiny, or glowing, appearance. It might even have been emitting some sort of shimmering, semi-visible field, which obscured any clear view. It was almost looking at the highway in the distance on a blisteringly hot summer's afternoon, when it seems like there are glimmering pools of reflective liquid in the distance, or maybe even suspended in the air. That's what it seemed like, anyway, when I looked at the spindle end-on.

When we pulled up to Walter's house, Merle's familiar vehicle was parked at the curb in front, waiting for us. "About time you showed up!" he told me, with a big smile on his face. "Did you have to stay late, or what?"

"No, I got out on time," I said. Truthfully, I practically ran out of there, to avoid more questions by Lillian. "I had to stop by the pet food store. Then I had some issues carrying all my stuff. Then the girls, uh, rescued me."

Merle smiled at the sight of me valiantly trying to pick up and carry all the loose items in the trunk. "Here, Ken, let's each grab a few things."

"Thanks, Merle."

Merle and all three girls each grabbed a handful of items to bring into the house. I had the bag of cat litter. We walked up the stairs, I opened the door, and we all went into the house.

We entered the kitchen and put the pet supplies on the counter. As soon as we set everything down, Atropha began to whistle in a light, high-pitched tone. "Here kitty kitty!" she said. I was surprised to see L.C. come into the room, and make a beeline for Atropha. Before long, Atropha was on the couch, and L.C. was on her lap, being petted and purring up a storm.

"Wow!" I said. "I've never heard L.C. purr so loudly, before."

Atropha looked like she was in heaven, with that cat on her lap. She was petting L.C. behind the ears, around the head and neck area, and all along her back, down to her tail. L.C. was eating it up. I was shocked, because I had never seen Atropha so peaceful and happy before. Usually she was a tad surly, if you know what I'm saying.

Merle noticed it too, and he gave her a little ribbing. "Atropha, you don't seem so fierce, anymore, with that little cat on your lap. I think I'm going to start calling you 'Sweet Atropha.'"

Atropha replied, without even looking up. "Well, Merle, maybe I'm starting to relax a little. It's nice, for a change, not being constantly accused of being a witch, and not having people trying to take my head off with a pole-axe or halberd every time I turn around!"

Merle laughed, at that comment. "Good point, Atropha! I suppose not being pursued by angry mobs does have its advantages."

I wasn't exactly sure what a "pole-axe" or a "halberd" was, and I really didn't have much idea what they were talking about, with "angry mobs" and all that.

Clotro walked over to where Atropha was sitting with L.C. "Hey, don't be such a cat hog, Atropha! I want some, too!"

"Me, too!" said Latsis.

"Too late," said Atropha. She carefully lifted L.C. off of her lap, and she stood up. "We have to go."

I don't think I've ever witnessed such an abrupt departure. Within a minute and a half, at the most, they were all out the door, in the car, and down the street and out of sight.

As I looked out the window and watched the little red car drive away, I mentioned to Merle that I was amazed at Atropha's affectionate behavior towards L.C.

"Well," Merle said, "Everybody likes a little affection, Ken. Even a hard-core inter- galactic time-saving security agent enjoys a little loving. Behind that gruff exterior is a real heart of gold, believe it or not. Imagine how lonely it must get for those time- savers, out here on this mission, so far from home. That cat just made her very happy."

"I see that now, I guess. But what the heck is a 'pole-axe', or a 'halberd'?"

"Oh, those are common types of medieval weapons. Remember, these ladies are time-savers. So to them, medieval times were maybe only a few months ago, or less. A group of ladies like them tended to encounter a lot of people that wanted to whack them in the head. And Atropha is the security agent for the group, so she would usually have to do something about it.

I shuddered as I remembered Atropha's "garden shears". "Why did people want to whack them in the head?"

"Why?" Merle laughed. "Well, for one thing, they liked to accuse the girls of being witches. You don't know very much about medieval times, do you?"

"In truth, no."

"Well, let's just say that getting whacked in the head by some medieval weapon was not an uncommon cause of injury and death, back then."

"Seriously?"

"Oh, yes. If you think the times you live in are violent, which they are of course, you'd have to see what it was like 500 years ago, or 1500 years ago. In general, violence was progressively more prevalent in daily life, the farther back you go. Of course, these days, weapons are far more powerful than stone axes, or pole-axes and halberds. That's a rather unfortunate equalizer, in the present day."

I pondered that thought for a few moments before a question popped into my mind. "Merle, why did the girls leave in such a hurry?"

"Why? Because your parents are coming."

"They are?"

Before Merle could answer, the doorbell rang. "Yes," said Merle. "They are."

I answered the door, and my parents came in. My mom was carrying a bag with leftovers from last night's dinner. They both said their "hellos", and then my father asked if I had any coffee.

"Sure. Want me to make some?"

"Heck yeah, I'd like some," he said.

I brewed up some coffee, sort of strong like my dad likes, and I poured a tomato juice for Merle, without him even asking. He winked at me when I offered the tomato juice, which I know was his way of thanking me for stocking the one Earth food or beverage that I knew he liked. We talked with my mom and dad for almost an hour, while we all took turns playing with L.C., who was in a very frisky mood after her great petting session with Atropha.

Eventually, my mom mentioned that they had to pick up some groceries so she could prepare dinner that night. That was all that my dad needed to hear, and they took their leave of us. I was relieved that the entire visit seemed to go quite well.

"Super nice people, your parents." Merle observed. "Yes."

Merle went to the window and watched them drive off. "OK," he said. "Time for us to hit the trail, also."

"So where are we going today?"

Merle placed his hand on my shoulder. "My friend, I don't think we will ever forget what we are going to do today, and where we are going to go today. I'm looking forward to it tremendously, myself."

# CHAPTER 34

Soon, we were back in Merle's car, and headed over to what was becoming the "usual" spot, behind the mall by the dumpster. When we got there, a truly scraggly- looking man was there, on his bicycle, peering into the dumpster. As we approached, he reached down into the dumpster and pulled out a long loaf of Italian bread. I noticed he already had several other items, apparently salvaged from the same dumpster, in the basket on the front of the bicycle.

Merle rolled down his window as we rolled up near the dumpster. "Sam, my friend!

How is it going today?"

For a moment, I was surprised that Merle knew this garbage-picker. For another moment, I thought that he might be an alien, just as I once suspected Professor Jonmur to be. Again, I was wrong. Merle simply had made a lot of connections, during the course of his visit, even extending to the local dumpster diver. After all, both Merle and he visited the same dumpster.

"Well, hello there, friend!" Sam said. "Good pickings today!"

Sam never did know Merle's name. According to Merle, Sam never could remember it. "I don't mind being called 'friend', anyhow," Merle told me.

After securing the loaf to the back of his bike with a bungee cord, Sam said "good- bye" and left the vicinity. Merle then reached next to his seat, and pulled out something in a small box. He opened the box. Inside was a watch. It was just a regular, inexpensive watch, from a local store—the price tag was still on the box. He handed it to me, saying, "Here, you take this watch. Put it on your wrist."

There was a time when this would have frightened me, in some way, but by now I trusted Merle implicitly, and put it on my wrist.

Then, Merle backed his car up a few yards, so that we could see the other side of the street, around the corner of the building. "See that clock?" he asked me. There was a large digital clock type sign in front of the bank that was there.

"Yes, I see it."

"What time does it say?"

"3:14."

"OK. Now what time does your watch say?"

"3:14."

"OK, good. The clocks are synchronized. Or close enough for our purposes." He pulled the car back forward, out of sight of the street. "Are you ready?"

"Sure."

"OK, here we go."

The darkening came again, and I noticed more clearly that it effectively obscured any details outside of our windows. It lasted for a few moments, this time, and then, suddenly, an incredible sight loomed before us. It took my breath away, and for almost half a minute I just sat and stared. I was too transfixed to look away, even for a moment.

"So, what do you think?" I finally heard Merle ask. I was so focused on the sight before me that Merle's disembodied voice seemed to seep slowly into my consciousness, as though through a fog, before I became consciously aware that Merle had even asked a question. Suddenly I jolted almost out of my seat, as our situation— perhaps I should say our position-- finally sank in on me.

"What do I think? What do I *think*? Is that what I think it is?"

"What do you think it is?"

I choked on the words, and they almost didn't make it out of my lips. I was feeling a lot of emotion at that moment. I think I sort of gutturally whispered my response.

"Saturn?"

"Yes! Saturn! You are correct, sir!"

It was undeniably true. In front of us loomed the magnificent planet Saturn, gleaming in a strangely subdued, yellow-gray froth, its spectacular giant rings surrounding it and shimmering like a scintillating halo of muted diamonds. I had been holding my breath, without realizing it, and now I let it out in a long, deep exhale. "Holy mother of pearl, Merle! I can't believe it! I just can't believe it."

"Well, believe it."

Merle let me sit there watching in silence, completely mesmerized, for maybe three more minutes or so. Eventually, I scanned around a bit more, and was alarmed to notice an extremely bright star, back off behind us to our left. I had never seen anything quite like it. It was too bright to look at directly, like the sun was, but it wasn't much larger than a regular star in the sky. "What the heck is that, Merle?"

"That? Don't you know?"

"I have no idea."

Merle laughed, quite loudly, at that. "You have no idea, huh? Don't you recognize your own sun?"

I looked more closely at the bright shining light, and realized that it was true. We were so far from the sun, and it appeared to be so small at that perspective, that I didn't even recognize it! Although it was still too bright to look at directly, the sun's light did seem to be a bit more subdued, at that distance, than it was on Earth. I looked around some more, and noticed that the stars were shining even more brightly, albeit without the twinkling effect, than they were that night in the Himalayas. "Merle, I can't believe how beautiful this all is!"

"The universe is a wonderful thing, isn't it?"

"Yes."

We sat there for several more minutes, as I looked around. I noticed a few of the many, many moons surrounding the incredible planet. Merle took us over for a closer look at the ethereal Titan, which looked not unlike a fuzzy, pale orange ball, with some slight variations in color density, and a thin blue atmospheric haze surrounding it.

"A lot of interesting things going on with this moon," Merle said. "Too bad we can't see very much, with this orange smog. Would you like to see some of what is going on below?"

"Sure I would!"

Merle tapped the screen in front of him, and a holographic-type image of the moon appeared in that spot. "Let's get rid of that smog," he said. Merle tapped the screen again, and the orange soup on the holographic view began to melt away, revealing a surprisingly complex landscape below.

"Are those oceans down there, Merle?"

"I don't know if 'oceans' is the right word. But there are large lakes down there."

"They are gorgeous!" The mirror-smooth lakes appeared to reflect diffuse sunlight, and starlight, from the surface, like silent ponds of mercury. "We probably wouldn't want to swim in those lakes, though, would we?"

"Ha! Not those lakes, Ken. They're not water! They'd peel your skin off, if you didn't freeze into a solid block first."

I wanted to stay longer, but Merle said we had to keep moving. Our next stop was the much, much smaller Enceladus, pale and ghostly white against the backdrop of blackness and dazzling stars. From that distance, we could clearly see a long jet of material spraying out into space from the tiny moon. "That's water, isn't it, Merle?" I remembered reading about Enceladus, so I knew about the water jets that issued from fissures in the fascinating little moon's crust.

"Yes. Actually ice, really, but it starts off as water. There is a lot of water on Enceladus, to say the least. This is a much younger moon than Titan."

"Wow," was all I could bring myself to say.

Merle actually took the ship right through the jet, and we could see diffuse streamlets of water ice swooshing across the front "windshield" of the ship as we passed through. The particles of ice were so tiny that it actually looked more like water, than ice. "Could we drink that water, Merle?"

"We could probably drink it, I imagine. But you'd probably still want to filter it in some way."

"Well, can we collect some of the water, then?"

"Absolutely not."

"Too bad." Right then, a tiny object, floating out some distance away, caught my attention. "Merle, is that what I think it is?"

"Yes, it is! Good sighting, Ken! Would you like a closer look?"

"Would I? Are you kidding me?"

"Hah! Here we go, then!"

In another moment, we were floating alongside the object- an interplanetary NASA probe, launched from Earth. I had goosebumps over my *entire* body, watching it going about its business. The probe had taken several years to travel out to that distance.

We made it there, quite literally, in moments. But I was concerned that we were exposing ourselves to its observation. "Aren't you worried it'll see us, Merle?"

"No, not at all. We're perfectly cloaked. Nice, isn't it?"

"Heck, yeah, it's nice! It's absolutely amazing!"

"It's wonderful to see a ship from Earth out here, that's for sure. Good for Earth!"

"Yes." I suddenly felt incredibly proud, that our plucky little probe had made it all the way out there!

Right then, Merle tapped the screen again. "OK, time to go," he said.

"So soon?" I probably whined, like a young child being pulled away from an amusement park.

"Oh, don't worry. We're not going back home yet."

He was right. After another few moments of darkening, another astonishing spectacle was staring us right in the face. This time, I knew what I was seeing, right away. "Neptune!" I gasped.

Once again, it was true. Once again, I was face-to-face with something I had only seen photos of before, something that seemed surreal, even unreal, at this close distance. The gem-like, azure color of Neptune was truly stunning, like the coolest blue cotton candy ever, swirling around in a perfectly round globe, streaked with high, white clouds, and scattered darker areas and streaks, as well. The planet seemed much smaller than Saturn. "Is Neptune smaller than Saturn, Merle?"

Merle looked at me, no doubt surprised that I would need to ask such a question. "Well, the diameter of Saturn is about nine times as large as Earth's diameter, I'd say, and its mass is probably about 95 times that of Earth. Neptune, on the other hand, is about four times the diameter of the Earth, and maybe 17 or 18 times as massive."

"Oh." Our ship was slowly rotating, and soon a moon came into view in front of us.

This moon was rather similar in appearance to Enceladus, but significantly larger. I noticed that it also was producing jets of material. There were actually several active jets, but they didn't seem to be blasting as strongly into space as the jets from Enceladus. "What's the name of this moon, Merle?"

"This is Triton."

"So we went to Saturn to see Titan, and now we're at Neptune, looking at *Triton*."

"Yes."

I actually was trying to make a sort of dumb joke over the similar names, but Merle didn't seem to get it. "Is that water, also, Merle?"

"No. You don't want to drink this stuff. It's liquid nitrogen, mostly."

"Oh." It was then that I noticed the rather subtle, and much darker, rings of Neptune.

There seemed to be four, maybe five rings. "Why are these rings so dark, Merle? I almost didn't even see them!"

"Saturn's rings are mostly ice, which is why they are so reflective. These rings of Neptune are mostly organic compounds, and dust. They are actually much more interesting, in many ways. They just might not be as beautiful."

"They're still very beautiful, in their own way."

"Yes, I think so, too. Come, we have one more stop."

"Merle, why are we always in such a hurry?"

"A hurry? It's not that we're in a 'hurry', necessarily, but that we are on a schedule."

"What kind of schedule?"

Merle went back to the windshield, and he touched the screen. He checked something out, tapped the screen again, and looked back at me. "OK. Let me summarize, quickly. Out here, there are many ships, from many different civilizations, as I've already mentioned, not only in orbit around Earth, but also out and about around the entire solar system, in stealth mode. When we do move about within the sphere of this solar system, our preferred mode of travel is at relativistic velocities. So we coordinate, with each other, to avoid crashing into each other at relativistic velocities.

All requests for travel are logged into a central server, as you might call it, and travel arrangements for all are tightly pre-scheduled. So when we have to go, we have to go."

"What would happen if we just--" I never got a chance to finish my question. I guess I know what the answer would have been, anyway, more or less.

"We have to go," said Merle. "Now." And again, the darkening happened.

Our visit to Saturn and Neptune had definitely taken me by surprise, even considering I never knew what Merle might do next, or where he might take us next. But the next stop was something that I probably might not have even considered as a possibility, even in my wildest dreams.

# CHAPTER
# 35

This time the darkening occurred for a noticeably longer period of time, but still only a few seconds, at most. We seemed to blink back into existence, parked in front of another large planet. But things seemed very different, this time. Everything was noticeably darker, with the exception of the still-blazing field of stars in all directions, and I noticed that the sun was even still noticeably smaller- no different in appearance than many of the other stars in the sky, except for its still-painful brightness. I pointed to it, to make sure I wasn't just confused. "Is that our sun, Merle?"

"Yes. We are much farther away, now."

"What planet *is* this, Merle?"

"It doesn't really have a name yet."

"What do you mean, it doesn't really have a name yet? Are we in a different solar system?"

Merle looked at me and blurted out a short little chuckle. "No, we're still in the same solar system, Ken. It's just that this planet hasn't actually been discovered yet. Well, not exactly, yet, anyhow."

The planet was a dusky color. It appeared to be a grayish, purplish blue, but I wasn't exactly sure what colors I was looking at, exactly, in the very dim light of the very distant sun. The surrounding starlight, I had noticed, seemed to imbue everything in a cool, even cold, light. "How far away from the sun are we, Merle?"

"Oh, I'd rather not say. I'm sure I'm not supposed to say."

"OK. It's far, though, I can see that."

"Yes."

Suddenly I remembered reading several articles about a very distant planet that had been theorized, due to its estimated gravitational effects on various other objects in the outer solar system. "Oh my gosh, is this the 'planet 9' that has been theorized? I've read about that!"

"Could be something like that," said Merle.

It was difficult for me to compare sizes of these various celestial objects, from the vantage point of the ship. "How big is this planet, then?"

"Well, it's bigger than Earth, I'll tell you that. But I'm sure I'm not supposed to say very much. I just thought you'd like to see it."

"I do. It's very cool, Merle."

"Cool." Merle enjoyed using slang in conversation, you could tell.

I thought that I saw some faint rings, and possibly a moon, also, but it was hard to see as clearly out there, so far from the sun. But suddenly I saw something else that grabbed my attention. There was a space ship out there, not very far from us. Except the ship seemed enormous- it completely dwarfed our own ship, like a large watermelon, compared to a grain of rice. Also, it was shaped more like a cigar- or maybe a gigantic grain of rice, if you will.

"Merle! There's another ship out here with us! And it's HUGE!"

Merle wasn't surprised, or alarmed, in the slightest. "Actually, there are several ships out here with us. The other ones are on the other side of the planet. Those are mother ships that are staying out here, getting a little research done while their scout ships fulfill their missions on Earth. Yes, they are rather large, compared to our little ship. In fact, they are very large compared to our base ship, as well. You could probably put about 50 of our base ships inside of this mother ship, quite literally."

"How in the world can anybody build something that large, Merle?"

Merle laughed. "Remember, Ken, just about everything in the universe is relative to something else. Maybe you won't be surprised to know that there are ships out there in the galaxy that are 50 times again as large as this ship we are looking at. And who knows, after that, you know?"

Again, Merle gave me some time to let it all soak in. After a while, a question popped into my mind.

"Merle?"

"Yes?"

"How is it going with the other contacts? I mean, the other two Earth people that your friends are communicating with."

"Not so good," said Merle. "What do you mean by that?"

"Well, each of my colleagues already went through three potential contacts, without success."

"So a total of six failed contacts?"

"Yes. Well, a total of eight, actually."

"Eight? I thought you said both colleagues went through three potential contacts."

"True. And I am on my third."

Well, that was a shocker, I have to tell you. "You are on your third? Am I the third?"

"Yes. You are the only contact left, out of nine. Congratulations."

"What happened?"

"Well, six of the eight insisted on continuing to try to take photos, one refused any attempts at communication, and one guy just sort of flipped out during a contact."

"Flipped out?"

"Yes. We had to cut him loose from the program, early on."

"What did he do?"

"For one thing, he physically attacked our team."

"Really?"

"Yes, really. I have the scar to prove it."

"Oh my gosh! He attacked you?"

Merle didn't want to talk about it anymore, and he waved off the entire line of questioning.

Well, I certainly didn't expect this development. I had always felt like I had a couple of teammates out there, that I might get to meet someday, if all went well. So hearing the news about the other eight potential contacts hit me like a ton of bricks, as they say. And knowing that I was the final fallback for Merle, as well, made it suddenly feel like a lot more pressure. But I didn't let on, too much, I don't think. Still, I could see that Merle was watching me closely.

# CHAPTER
# 36

It seemed to me that Merle went out of his way to quickly change the subject. "This is as good a spot as any to talk a little Physics," said Merle.

"Sure." I think I was glad to change the subject, anyhow.

"We've already discussed the 'gravitational soup', or 'dark matter', which results from all the extra-dimensional matter in the universe, and its effects on the gravitational coherence of galaxies, and that sort of thing. But that gravitational soup is a gradated soup, with higher areas of gravitation closest to hyper-dimensional galaxies and hyper-dimensional galaxy clusters."

"OK."

"Do you think that there might be a different sort of gravitational soup-- a perfectly non-gradated gravitational field, identical and not curved or compressed in any direction— which might have some effect on fundamental particles at the quantum level, as well?"

Every once in a while, Merle prompted me to think about something, and I instantly saw what he was getting at. This was one of those instances. "The Higgs field!" was all I could say, as my mind began to race.

The "Higgs particle", or "Higgs boson", was named after Peter Higgs, one of a group of six physicists which first postulated its existence back in 1964. The particle was said to be a quantum excitation of the theorized "Higgs field," an overall field which exists throughout the entire universe, and allows fundamental particles to gain traction, as it were, and thereby acquire mass. The whole idea of the field, and exactly what it represents, has been sort of a Holy Grail of physics ever since. I immediately realized that this perfectly non-gradated "gravitational soup" that Merle referenced might well fulfill many of the needed parameters of the long-sought field, so essential to our entire quantum-based existence.

"So where does this non-gradated soup come from?" I asked Merle.

"We will get to the non-gradated soup soon. First I wanted to discuss its effect on fundamental particles."

Again, goosebumps broke out over my entire body, as I sat there, before a good-sized, somewhat undiscovered planet, within sight of a gigantic, mega-scale mothership, and almost casually discussing the solution to a very long-held problem in the field of particle physics. "Are photons affected by the soup, Merle?"

"Well, not so much, Ken. Photons simply ride the wave of the space/time continuum, with basically little or no resistance. Their energy can be absorbed by electrons, but they are not really able to gain traction within the soup, to speak of. That's one of the special things about them. But other fundamental particles don't surf space/time in the same way. Those particles are able to link up with other fundamental particles, due to their carrier particles which are enabled by the powerful non-gradated gravitational field which is provided by the extra-dimensional mass/energy we will soon discuss. That's how these particles are able to gain a foothold, you might say. This field is why physical objects have inertia."

"How do the particles link up?"

"Well, do you have any ideas of their shapes?"

"No. How could I?"

"Do you know the shape of any other fundamental particle?"

"Just the photon", I said.

"And what is its shape?"

"Well, it's a loop. Or a perpendicular pair of loops, if you're breaking apart the two components."

"So what about the shape of other fundamental particles?"

"Are they loops, as well?"

"Of course they are, to a certain extent, at least."

"Oh." Again, as so many times before, this revelation seemed fairly obvious, as I digested it for a while, before Merle asked another question.

"So, Ken, how do you think they link up, then?"

"I have no idea."

"Think about it some, Ken."

I gave it a little thought, and suddenly the idea burst right into my consciousness. "Oh! Are they interlocked? Like one ring, linked through another ring?"

Merle gave me that smile he had, whenever I had a huge realization. "Generally speaking, Ken, yes. Each particle describes its own type of ring, and these rings are able to interlock together, or knot themselves together, in very specific ways-- in a vast variety of very specific ways, mediated by their carrier particles, which represent the strong and weak forces of the Standard Model."

I tried my best to get more information out of him, since I was greatly interested in this line of thinking, but he just winked at me. "That's all I'm going to say about fundamental particles for now, Ken. You're on the right track, now. Everything else will fall into place, eventually, as the centuries go by."

"Centuries!"

"OK. 'Millennia' would probably be more accurate. Or, most accurately, there's always more to learn. Do you think the universe reveals everything all at once? It never does that. These things take time."

Suddenly a sound emanated from somewhere on the ship, and Merle touched the screen in front. A message, in English, appeared in front of us, hovering in mid-air like a glowing, two-dimensional hologram. It said, *"Welcome, Merle and Ken, to the outskirts of this solar system! Are you enjoying your tour?"*

I stared at the welcoming message in wonderment. "Who in the heck sent that, Merle?"

Merle pointed to the giant mothership. "Our friends over there sent it! How do you want to respond?"

I thought about it for a moment. I didn't even question how the inhabitants of the mother ship knew who we were, and how they knew what were we doing, or even who they were, or where they were from. It was all simply par for the course. "How does this sound? '*We are having the time of our lives out here. It is all very beautiful and wonderful.*"

"Sounds great," said Merle. "I'll send it." He tapped the screen a few times, and sat back for the response.

Moments later, the response appeared in the same spot and same way as the previous message. It read, "*Yes, it is wonderful. We are also having the time of our lives out here!*"

Merle looked at me again. "Response?"

"*It is very nice to meet you.*"

"I like it!" Merle said. He sent the message, and we again waited for a reply.

It took a few more seconds this time, but soon enough the new reply appeared in the same spot. "*It is very nice to meet you, also! We wish you both the best of luck, and the greatest success, in your endeavors. With warmest regards, your friends, 'from out there.'*"

"I think I can handle this one," said Merle. "*Thank you for the kind regards, and best wishes for your own safe journeys. Your friends forever, Ken and Merle.*" Merle sent the message, and spun back towards me. "We have to go. We can't stay long at this next spot. Eighteen seconds at most." And the brief darkening happened again.

This was the longest period of darkening, by far, and after about ten seconds I asked Merle if everything was OK.

"Yes, everything's fine."

Just then, we were suddenly face-to-face with another planet-another gas giant, and an especially colorful one, with a gigantic ring system, noticeably larger than Saturn's rings.

"Where are we, Merle?"

"We've accelerated into another physical dimension, Ken. Right now, even though it feels like we're standing still, we're traveling twice the speed of light, compared to our original position. All you see here before us is invisible to anybody back on Earth, just as we could never see Earth from this perspective."

"So this entire galaxy is traveling twice the speed of light, compared to Earth?"

"Well, this is a portion of the *Milky Way* that is traveling twice the speed of light, compared to Earth."

"Wow."

Another 15 seconds or so of darkness ensued, and abruptly, there before us, was yet another planet. This planet was stunning in its bright, radiant colors, massive oceans, and interesting landmasses. It was partially obscured, yet visually enhanced, by a multitude of fluffy clouds. We were back at Earth, and it was stunning. For the first time, it occurred to me that the most beautiful planet in the system was not Mars, or Jupiter, or Saturn, or any of the other planets. They all paled in beauty, compared with Earth, and I completely shifted my focus to the planet.

"Oh my gosh, Merle. It's so awesome! Is that Australia, and Asia right down there?"

"Yes. All in all, Earth might be the most beautiful planet I've ever seen. Although, I have to tell you, Akeethera isn't too bad, either."

I smiled at that. "I'm sure it's very beautiful, Merle."

Merle touched the screen, and up came a picture- a hologram, I guess, or something of that nature. It was a planet that looked surprisingly similar to Earth, with perhaps fewer oceans but with more green, and three small, fairly nondescript moons in orbit around it. Merle reached out and zoomed in for a closer look at the planet. "Here it is. What do you think?"

"That's Akeethera?"

"Yes."

"It's amazing, Merle. It's so green! I love it!"

This time it was my turn to allow Merle some time to soak it in, as I could tell he was enjoying the moment. The image began to slowly rotate, revealing the entirety of the planet. He gazed at it some more, and then touched the screen again, and the image disappeared. He looked just a little bit emotional.

"You must miss your home, Merle."

"Yes, I do miss it, very much. But this trip was the opportunity of a lifetime. I don't have a single regret, if only for all the great friends I've gotten the chance to know, in person." He patted me on the shoulder when he said that.

I left him with his thoughts for a moment, before asking him a question that I had wanted to ask for the last several days. "Merle, just exactly how do these ships fly?"

"I've been wondering when you were going to ask that! Well, I'm not allowed to tell you 'just exactly', but I'll be happy to explain it in a general manner. It works by creating a powerful, directional magnetic field. By doing that, movement along the central axis can be quite energetic. The ship simply is pushed along, with very little resistance, on the magnetic field lines. In that way, the ship is able to accelerate to hyper-dimensional velocities almost instantaneously. It's sort of like a spinning bead on a string, I suppose you could say."

"So how is this magnetic field generated?"

"Basically, the engine is a torus of a super-conductive super-fluid, rotating at relativistic velocities. This generates a magnetic field, in much the same way a neutron star or black hole generates its magnetic field. Except the black hole's field is exactly the same to the north and the south, which stabilizes its position. On our ships, we are able to manipulate the directionality of our ship, so we can draw the ship in any direction we want, and change directions instantly. A ship can have any number of magnetic torus engines, depending on its size and design. Our little ship here has one primary torus in the center, and two sets of three additional tori surrounding it. That's a common set up for a lot of small ships, while larger ships, like our base triangle, have a lot more. I think our base ship has sixteen total engines.

"And why the triangular shape of the base ship?"

"Oh, there are various reasons for that, really, like flight stability and tori arrangement. One big reason is that we are able to link multiple triangular ships together, especially for long inter-galactic journeys, much like slices of a pie. Our base ship originally came out here as one of a group of six linked ships, which basically formed a giant disc. Or a giant hexagon, I suppose."

"Wow, that's pretty neat! Six slices of pie!"

"Yes, that's it, exactly! And now I have a question for you, to get back on the subject of fundamental particles."

"What is that?"

"Well, you now have a good idea of how our universe works, don't you?"

"Yes."

"Although the universe is actually *infinite* in terms of dimensions of time, you can see that the path of mass-energy in our universe is actually *finite*, eventually returning back to the center, and then repeating the process."

"Yes. It almost sounds like a paradox, but it's not."

"True. So, if space/time itself is infinite, which it is, what do you think might be outside of our universe?"

I sat there for several moments, thinking about that question. I had never even considered the idea, previously, but I could see what Merle was getting at. "I don't know, Merle. Infinite nothingness?" That sounded horrible, as I said it.

"No, I highly doubt that. Listen to my question again. I'm giving you a sort of clue, in the question."

"OK, ask it again, then."

"What do you think might be outside of *our* universe?"

Merle's heavy emphasis on '*our*' made it a lot easier, this time. "Oh. Not *our* universe, then. But it might be *their* universe? Is it another universe, Merle?"

"That's sort of it."

I gave it some more thought. "Is it a whole bunch of other universes, Merle?"

"You're getting warmer."

"Is it an *infinite* number of universes, Merle?" I was just guessing, at this point. But all roads seemed to point towards infinity, in the universe, so it seemed like a fairly safe guess.

"Yes. An infinite number of universes, indeed. On Akeethera, we absolutely believe that there are an infinite number of universes."

"How can that be?"

"Well, we believe that this infinite number of universes exists in the form of a simple cubic crystal. If you look at a simple cubic crystal, you'll find that each unit of the crystal is surrounded most closely by six other units, with another 12 units somewhat farther away. In this case, each unit is actually a universe. That's why, on Akeethera, we refer to the greater universe as the 'Infinite Crystal.'"

"The Infinite Crystal? Wow. That sounds like a piece of jewelry, or something!" Elegant title or not, my head was felt like it was about to explode, as I considered this new idea. Merle was placing our infinite universe within an infinite field of additional infinite universes!

Merle continued with his description of the Infinite Crystal. "So *our* universe is infinite in terms of dimensions of time, since there is always mass/energy that is traveling faster, or slower, than whatever mass/energy you might consider. The Infinite Crystal is infinite in terms of three-dimensional space, as well as in dimensions of time. You can travel through this field of universes forever, and there will always be more universes.

It's neatly arranged universes without end!"

"But how can there be no end?"

"How can there *be* an end?"

"What do you mean?"

"I mean, if you're looking for an end to the universe, maybe you're looking for a giant wall, or something. Maybe somebody built a giant wall, or something, blocking us off from going further?"

"I guess that wouldn't really make sense."

"There's no need for the space/time continuum to have any end, Ken. Just like there's no need for it to have ever had a beginning. There was always a yesterday, there will always be tomorrow, and there will always be an open road, in whichever direction we choose to move in."

"I guess that makes sense."

"Now here's another interesting part of all this," said Merle. "Perhaps you've wondered why our universe contains so much matter, yet so little anti-matter."

"Sure. Every astrophysicist wonders about that, I think."

"Imagine our universe, comprised primarily of matter, surrounded most closely by six other universes in the Infinite Crystal, and then another twelve somewhat farther away. Like polonium."

"OK. Wait, did you say like *polonium*?"

"That's right. Polonium, as you call it here on Earth, has a similar structure. Now, *our* universe is comprised primarily of matter, or at least what we call matter, in our universe. The six other universes that border us most closely, in the Infinite Crystal, are comprised primarily of *anti-matter*, from our way of looking at it. The twelve universes that are somewhat farther away are, again, comprised primarily of *matter*. And that pattern repeats itself throughout the Infinite Crystal. In that way, matter and anti-matter are able to exist in equal amounts, yet still avoid meeting up with each other, for the most part, since mass/energy does not cross between universes under normal circumstances. In a gravitational sense, the continuum doesn't care if something is matter, or anti-matter. It's all the same, to the space/time continuum."

Merle didn't have to tell me that, when matter meets anti-matter, a catastrophic annihilation of both the matter and anti-matter occurs. "So the symmetry of the greater universe is maintained, in terms of equal amounts of matter and anti-matter, and yet the entire system doesn't just go up in a big puff of annihilation," I told him, surprised at the neat solution to this previously vexing dilemma.

"That's right."

"Merle! That's like the secret of the universe, right there!"

Merle smiled at me again, like a parent whose child just discovered some fundamental aspect of existence that every adult already knows. "I think you've said that before, Ken."

"And I'll bet you'll tell me that the universe holds no secrets!"

"Yes."

"But if there is mass/energy traveling faster and faster and faster, won't that mass/energy be traveling farther and farther, and eventually bleed over into the adjoining universes?"

"No. Because four dimensional space curls in upon itself as the reference frame shifts away. That's why on Akeethera we refer to the physical space/time reference frame as a 'space within a space within a space'.

"I'm not exactly sure what that means."

"Well, one thing it means is that mass/energy of any velocity circuits back toward its universal black hole, before it has a chance to bleed into an adjoining universe. I guess you might say that it is the basic nature of the Infinite Crystal. That's why there is equilibrium between adjacent universes, with empty space in-between, just like the empty space in-between atoms of a crystal. There is enough space between universes to avoid any bleed-over, and yet they are close enough together to maintain their gravitational attraction in a crystal arrangement."

"So there is an outer edge to our universe? What happens if you're in a galaxy that is close to the edge?"

"Well, under normal circumstances, when mass/energy is traveling within the normal stream of the universal flow, there really is no such thing as a perceptible 'outer edge' to our universe. Remember that, from any possible location in our general area of the universal flow, galaxies traveling away at relativistic velocities surround the location, on all sides. So, from inside the galaxy, you can never really be aware of being near an 'outer edge', unless, possibly, you can accelerate to a fast enough velocity, perpendicular to the flow, and straight away from the center of the universe. That's one of the great mysteries—what would happen if you could escape the gravitational influence of the universe, and travel to that area of the Infinite Crystal that is in between universes? The problem is, not only would you have to escape the influence of our universe, but you'd also have to contend with the gravitational effects of the Infinite Crystal itself. Many people believe that it would not be possible to escape any given universe, under any circumstances. Some believe it can be done, however."

"But wouldn't the gravitational attraction from the other universes actually make it easier to escape our universe?"

"One might think so, but the true situation is probably the exact opposite. Let's take a look at how that works. To me, possibly the most elegant aspect of the greater universe is how the Infinite Crystal, this most massive of mega-large scale quantum- based structure, affects fundamental quantum matter at the tiniest of scales, via the non-quantum medium of the space/time continuum."

"How is that?"

"Well, again, space/time is an infinite continuum. And a universe is the ultimate warping agent of space/time. So each universe in the Infinite Crystal creates a powerful gravitational field, outside of itself. Now how do you think that might affect *other* universes?"

"Are you saying that space/time in our universe is gravitationally warped by the mass-energy in all the other universes in the Infinite Crystal?"

"I am."

"So *that* is why the field that allows fundamental particles to acquire mass is so perfectly static, throughout the universe. It's the cumulative gravitational field generated by all the other universes in the infinite, perfectly crystalline greater universe, so it results in a perfectly non-gradated gravitational field! As people, we don't even notice it, since we are far too large to even notice non-gradated gravity! We only notice gravitational *gradients*! Non-gradated gravity doesn't push us in one direction or another, so we don't notice it, as people. It's still plenty powerful enough to bog down force-carrying particles, though, and other fundamental quantum units! That's what allows them to gain mass and form aggregate structures! And also, non-gradated gravitation describes the perfect medium to result in the concept of inertia!"

"Bingo again, Ken. Although you can't really say we haven't "noticed" this non- gradated field. We've been looking for the source of the "Higg's field" for some time." Merle looked downright pleased at the roll I was on. "Otherwise, it's all as you describe. The field generated by the Infinite Crystal is identical and non-gradated, wherever you might travel in the universe. We can really only measure its strength indirectly, because we know the field must exist, or fundamental particles could never accumulate mass, or join together into larger, stable structures. The portion of the field that we cannot otherwise account for has to be from the gravitational effects of the greater universe."

"So the cumulative gravitational effect of the greater universe, aka the Infinite Crystal, is really what creates the overall Higgs field that permeates the entire universe?"

"To the largest extent, if we are simplifying it somewhat. It also permeates every *other* universe, as well."

"Yowza! That is huge, Merle! And it's small, at the same time!" I amused myself with that comment, while Merle didn't seem to even notice.

"Also," Merle said, "this field holds each universe in place within the Infinite Crystal, since each universe is being pulled equally in all directions, essentially, by the non- gradated field. This force is so strong and unyielding that the Infinite Crystal is the most perfect crystal that exists, anywhere, if you take scale into account. Also, that powerful force which maintains the crystal means that you'd really be swimming upstream, trying to leave your own universe." At that, Merle tapped the screen and looked over at me. "Say, Ken, what time do you have?"

"The time? Oh, that's right, my watch! It's 3:51. Wow, all that only took 37 minutes?"

"Well, the flying part went pretty quickly, didn't it?"

"I'll say it did!"

And with that, we took another brief look at the Earth. Then, another brief darkening occurred. Then it got slightly brighter, but still, everything was quite dark. "Where are we, Merle?"

"Back by the dumpster."

"But why is it so dark?"

"Because it's nighttime."

"But it's only 3:51!" I looked at my watch. "Excuse me, now its 3:52. But it doesn't get dark here at 3:52!"

"That's because it's not 3:52, here, Ken." Merle backed up the vehicle, turned the corner and headed for the street. There, in front of us, was the big digital sign.

"9:45! But how is that possible?"

"That's hyper-dimensional time dilation, Ken."

"Wow!"

"For us, in our dimension of space/time, as we traveled so quickly to the various planets and moons, only 37 minutes passed. Meanwhile, down here, it was six and a half hours. That's some hyper-dimensional time dilation, right there!"

"That's amazing!"

"Yes."

# CHAPTER 37

Merle drove me back home. He turned on his "radio" and we listened to some jazz. "Who is this playing, Merle?"

"That's Thelonious Monk."

"I like it."

"I *love* it," Merle said.

When we pulled up to the house, the fireflies were out, in the yards and in the park across the street. They were flashing all around us, and I could see that Merle was entranced by the spectacle. That was the night we sat there, watching the fireflies and listening to "'Round Midnight". For a few minutes, I think we both almost forgot all that we had experienced that day, and we just wallowed in the serene beauty and true marvel of nature that are the fireflies, with the music serving as a sublime aural backdrop. When the song ended, Merle touched the screen, and we listened to the entire song again, as we watched the flickering show.

"More beautiful than Saturn, I'd say," said Merle.

"I actually have to agree with you, Merle. I never really paid the fireflies that much attention, before, except maybe when I was very young, when we collected them in jars."

I think Merle may have winced a bit, when I mentioned collecting them in jars. But that's what we used to do, when we were kids. These days, with heavy pesticide usage, frequent grass cutting, and less leaf litter, which the underground firefly grubs require, there aren't nearly as many fireflies as there were in the past, but that night, at least, they were out in a decent abundance.

Merle and I said our "goodnight", and I made my way back into the house. L.C. came to greet me at the door, which was nice. "How is it going, L.C.?" I asked her. "It looks like it's just you and me tonight, girl!" I gave her some wet food and cleaned the waste out of her litter box. Then I listened to a voice message from Kim, and sent her a text message in return. It was too late, at that point, and I was too tired anyhow, to get together. By the time I finished replying to Kim, L.C. was sound asleep on her little cat bed, satisfied with a full belly.

"Correction," I said to her as she slept. "It looks like it's just me." I grabbed a quick bite to eat from the leftovers my mom had brought over that morning, and I plopped down into the couch, now totally exhausted. I quickly fell asleep, before I could even turn on the TV or the computer, and before I could even make it back to bed.

That night, I dreamed that Merle let me take his ship out for a flight, by myself. He assured me that I would be able to do it by myself, and I successfully traveled back out to the mysterious "Planet 9." I was hoping to see our new-found friends in their mothership, but they were gone by the time I got there. Panic set in when I attempted to make the return trip back. Instead of ending up back by the dumpster, I found myself in deep space, surrounded by stars on all sides, with nary a planet to be seen, anywhere. Nor did I recognize the sun, anymore. It was just billions upon billions of unreachable stars, in all directions. Every time I moved the ship again, in my attempts to get back to Earth, it seemed like the stars were just farther and farther away, and I realized that I was stranded, all alone, and likely for all time, in the vast desolate emptiness of infinite space/time.

I suppose I hadn't consciously realized it yet, but pressures and anxieties were building up in me. As much as I was enjoying having L.C. as my housemate, I knew that as soon as Neddie was released from the hospital, L.C. would move to the house next door, and I'd probably never see her again. I'm sure I feared I would lose Kim forever, also, when L.C. left.

On top of that, Merle was not long for this world, I knew for a fact. My impending mission of informing the world about hyper-dimensional relativity, and all the other truths I had come to learn about the universe, was weighing upon me quite heavily. It felt like the full weight of our universe had settled squarely upon my shoulders. I alone had to do what I knew would be a very, very difficult task, in convincing a skeptical Physics academia of the realities of the hyper-dimensional universe. I couldn't help feeling extremely isolated, and hopelessly insignificant; the tiniest and least significant little fish within the endless pond that was The Infinite Crystal.

# CHAPTER
# 38

It had been a very restless sleep, overall, as I endured an entire series of disturbing dreams. The dawn streamed in through the windows, and fell upon me, there on the couch, like thunder. Exhausted though I was, I got up, fed L.C., and made myself some coffee. After I took a shower and got dressed, I sat back down on the couch and wondered what this new day was going to bring. I didn't have to work that day, which was a Tuesday, and I found myself fearing that Merle had already left the planet, without saying goodbye.

My fears were assuaged, though, when I heard the familiar low tone, followed by the ringing of my doorbell. It was Merle!

"Merle! You're still here!" I told him.

"Yes, I believe I am," he said. "We still have a few more things to accomplish, Ken.

You're not going to be rid of me quite that easily."

"Good."

"Well, I'm glad you feel that way." L.C. came up to Merle, and rubbed up against his pants leg, purring. "Well hello there, L.C.!" Merle said. He reached down to give her a nice little petting. "Sorry to say, L.C., but Ken and I have to hit the road again."

"So where are you taking us today? Andromeda?"

"Now *that* would be quite a trip! But I'm afraid that might be a little outside of the scope of this mission. Today is going to be a little more down to earth, you might say. We'll be paying a visit to our friend, Professor Jonmur."

We got back in the car, and Merle turned on the radio. "What are we listening to today, Merle?"

"Well, right now we're listening to The Grateful Dead. The song is 'Ripple.' Do you know it?"

"No. I never really listened to The Dead, much. I guess I know a couple of their songs."

"Well, this one is easily my favorite. It really has quite a lot to say. It actually would make a heck of a good theme song for this mission."

We listened to the song as we drove along to the professor's house. I didn't really understand what it was they were trying to say, and I thought maybe Merle was just pulling my leg about it being a good theme song. I could see that Merle was into it, though, listening to the song and smiling, so we just drove along without saying anything. We pulled over and parked along a curb in front of a big green house that I assumed was the professor's, just as the song was wrapping up.

"Here we are!" said Merle, and we got out of the car.

The professor was out in front, puttering around underneath a large bush, and he turned to greet us when he heard us approach. "Well, hello, Merle! Hello, Ken! I'm so glad you both could stop by! I've dug out some dusty old things I thought you might be very interested in!"

"That's why we're here!" said Merle.

"Well, come on in, then. I have some very good tomato juice, if you're interested, Merle!"

"That sounds great, Professor!"

All three of us had a nice cold glass of tomato juice. "This is very good," I said. "I hadn't realized how much I do enjoy tomato juice, myself."

"They tell me it's an acquired taste, but I love it," Merle said.

"You always have, since I've known you!" the professor responded. "I've never seen you drink anything else!"

I sure wasn't about to say anything about Merle's unusual tomato juice habit.

Instead, I noticed several dozen frames hanging on the wall, across the way, in the den. Inside of each frame was a spectacular insect- either a large beetle or a butterfly, primarily. "Wow, are those the coolest insects, professor! Where did you get them?"

"I collected the majority of them back when I was a young entomologist. Ever since I realized how many fewer insects I was seeing, compared to when I was a younger man, I mostly changed over to just taking photographs. Catch and release, you know."

Merle and I both nodded in response.

"Now I have something else to show you," said the professor. He walked across the room and picked up a large portfolio that was leaning against a cabinet. "This is something I came across the other day, when I was looking through my father's old files. I hadn't seen these for many, many years. I was hoping that you'd be interested, Ken."

The professor brought the portfolio over to the table where we sat, while Merle cleared the table. The professor opened up the portfolio, and pulled out what appeared to be a set of large blueprints, or plans, faded significantly with the passing of time. He spread one plan out on the table, as Merle rejoined us. At the top of the large sheet was the title- "Maximillian R. Jonmur Streamside Park".

I sat up in surprise. "Is this the original blueprint for the park?"

"It sure is," said the professor. "Actually, there are several plans in the set. And I found a few interesting things."

"Like what?"

"For one thing, the factory buildings on the north side of the stream were supposed to be connected to a drainage system on the opposite side of the buildings, which flowed *away* from the stream, to avoid backflows into the stream during storms, as it does now. There was supposed to be an entire filtration system that was never built. Instead, several companies just ran some outflow pipes right down towards the stream."

"We've seen that," said Merle. "It's outrageous."

"You've seen that? How did you get back there?" the professor asked. "It's not easy to get back into that area, unless you wade down the stream, which would be pretty dangerous with the current, and all the garbage in there. I wouldn't recommend wading in that water without the proper equipment."

I thought I'd better interject. "We didn't wade out there; we just sort of crawled through the underbrush to check it out. I think I still have the scars to prove it!" Merle gave me a wink after I said that.

"Well, I'm impressed," said the professor. "You're more dedicated to this, already, than I've realized. Very few people ever make it back into that area!

"Another interesting thing I found was that my father had designed a fishing pond, as an oxbow off of the main body of the stream, so people could easily fish there without the current of the stream being such a factor. Also, he had designed an overflow outlet, so that when the stream flooded, some of the water would divert into a retention pond, with a recreational area surrounding it, across the street from where the factories are now. That area was zoned to be left undeveloped, in case it could be used for that purpose in the future. It runs all the way to the back side of the forest preserve, so it creates another corridor for wildlife, all the way back to the lake area. It's a swampy area, which is probably why they still haven't changed the zoning designation for that spot. So the site is still available, basically."

Merle moved in closer, and peered deeply at the plans. "Fascinating!" he said. "That's great news!"

"I know, right?" the Professor said. He pointed back to the plan. "And here's where my father had designed three underpasses, under the road, to allow animals to pass under. Notice here," said the Professor, pointing at an area of the map, "where he wrote that the underpasses should be constructed where the natural washouts are from the flood plain. Look, he even wrote, 'enhance the natural washouts', right there. He thought the underpasses should be like large, enhanced natural washouts that animals naturally gravitate towards, anyway. I remember him telling me, when I was little, that a lot of animals were being struck by cars on this road, so I know it bothered him that they didn't put in the underpasses."

"The underpasses are a great idea," said Merle. "That road is pure carnage, to this day. Not to mention, it's dangerous for any driver that hits a large animal. I recall a few years back, when a woman was seriously injured after her car hit a deer on that road."

"I remember that!" said the professor. "That was tragic. It could have been easily avoided with a simple underpass. In fact—" the professor paused for a moment to look at the blueprints. "In fact, if I recall, that accident was very close to where my father proposed one of the overpasses." He paused to consider the thought. "Well, I suppose I have some big dreams that may or may not be realized, some day. But first we have to get that stream cleaned up! In the meantime, I have something for you both." He reached back down into the portfolio, and brought out more blueprints that looked much newer. There were two rolls, secured with rubber bands. "I had three full sets of copies made," he said. "There is a set for me, so I don't have to keep using the originals, and another set for each of you." He handed us each a rolled-up set of copies, which we gladly accepted.

# CHAPTER 39

Merle was extremely happy as we got back into the car, and he turned on the radio to a very up-tempo instrumental. I believe it was the "Black Mountain Rag", in the country or bluegrass vein, with fiddles, banjos and guitars. Merle said it just "sounded happy" to him. "Love that Doc Watson, anyhow," Merle said. "Fabulous person, amazing musician!"

Merle looked over at me as if he was telling me a little secret. "Frankly, I was thinking that 'Big Yellow Taxi' would probably be a much better fit, but I was in the mood for something a little less wistful," Merle said.

"This is good," I said helpfully. I had no idea what he meant by his "Big Yellow Taxi" reference.

"Well, I'm glad you like it."

"So, what's next?" It was still early in the day, and usually, with Merle, there were places to go. This time Merle had a little surprise for me.

"What's next is, you are going to get a little bit of a break from me, and astrophysics, and all that."

"Why's that, Merle?"

"Well, they say that 'all work and no play makes Jack a dull boy', which is very well put. And lately, you've been all work, and no play. You need some time off for some 'R and R'. Also, you don't want to completely forget about your friends, do you? They are very important. I'm going to take you back home, and then you'll have the rest of the day free, as well as tomorrow. I'll see you on the following day."

This was completely unexpected, but I was surprised at how much it did feel like a big relief. I realized that, again, Merle knew what he was talking about.

Not five minutes after I got back home, my phone rang. It was Kim, calling me on her lunch break. We agreed to get together for dinner, after she got off work.

Not long after I hung up with Kim, the phone rang again. It was Ronny. "Hey, buddy, you're a hard man to get ahold of, these days!" he said. "You're not even answering your text messages!"

"I've been a little busy, I guess," I said.

"Oh, sure, you've got a girlfriend now, and you've forgotten all about us, haven't you?"

I had no good response for that one.

"The reason I'm calling," Ronny said, "is that I have three tickets to tomorrow's game, and Keith can't go. I doubt very much that Bryce would want to go, because it's Cubs versus Mets. I thought that maybe you'd want to bring your new squeeze. It's a 7:00 start."

"That would actually be perfect. I have to work the lunch shift tomorrow, so that will work for me. I can call Kim back and see if she'd like to go."

"Call her back, huh? OK, Romeo, go call her 'again', and let me know. Otherwise maybe we can find somebody else."

I talked to Kim, and it worked out perfectly for her, also, which was super encouraging, so I called Ronny back to confirm our plans for Wednesday evening. I still had some time to clean the house, and to take care of some garden maintenance in both Walter's and Neddie's yards, and to take another shower, before Kim came over. She took me out to her favorite restaurant—a Thai place-- where we had a nice, leisurely dinner, with two pots of tea, and dessert, as well. We came back to Walter's place afterwards, so Kim could spend some time with L.C.

"You really do like this cat, don't you?" she asked me.

"Yes. I'm going to miss her when she goes back to Neddie."

"Well, I have some good news I forgot to tell you in the restaurant. I talked with Neddie yesterday."

"So what's the good news? Is she getting out of the hospital?"

"She should be out sometime next week, apparently. But that's not the only good news I was talking about."

"So what's the other good news?"

"Neddie says that you can keep L.C. for as long as you're staying here this summer, if you'd like. That will give Neddie a little more time to recover, before she takes a pet. She also told me that she's planning to sell the house, probably within the next two or three years. One of my aunts has offered to let Neddie stay with her, until Neddie is ready to move into some type of assisted living facility or something, down the road.

After her bad fall, she realized that she is getting up in the years and could probably use some help with things, before she has an even worse accident. She's almost thirty years older than my mom, and lives by herself, so it can be tough for her at times.

"Neddie hopes that, once she moves in with my aunt, one of us can look after L.C. Neddie says that you're an absolute angel for helping to rescue L.C., and that you should have first option of taking her in. Neddie's only stipulation is that you let her stop by for visits, from time to time."

"Well, I don't quite know what to say. Other than it all sounds great. I'm not even sure where I'll be in two or three years, but it all sounds great!"

"I think it *will* be great," said Kim. And that's when she reached out and gave me a big kiss on my cheek. The next one was on the lips, and so was the one after that.

Let's just say we had a very nice time together that night! By the time Kim left, several hours later, it was obvious that our relationship had evolved from simple friendship into a much deeper, romantic relationship. I was on Cloud Nine, as they say, when I went to bed that night, and I slept like a baby, with no disturbing dreams, for a change.

Wednesday, Kim and I both worked during the day, and that night we had a tremendous time at the baseball game with Ronny, even though the Cubs lost to the Mets, 6-3. The three of us went out for some late-night pizza and a lot of laughs, afterwards. I crawled back into bed rather late again, tired, but also quite satisfied after a very enjoyable night out with two great friends. I slept hard, and soundly, again that night.

# CHAPTER 40

I had set my alarm the night before to get up early, since I knew I would be getting together with Merle on Thursday. Even on short sleep, I leaped out of bed, showered, and made coffee in record time. I took care of L.C.'s food and litter box, and after she ate and groomed herself, I sat down on the couch to play with her. I had a thick piece of yarn, maybe four feet long, and I tied a large goose feather that I had found to the end of it. I flipped the yarn this way and that, and L.C. was wild for it. She scooted back and forth, grabbing the feather in her paws and in her teeth, and she tried to run off with it. I was laughing at her like a crazy man, when the doorbell rang. It was Merle, of course. I should have known something was different, because I never heard the droning tone this time. But maybe that's because I was laughing so loud.

"Come on in, Merle!"

"What's going on? It sounds like you're having a party in here!"

"Oh, L.C. and I were playing with a piece of yarn and a feather."

"Well, like I said, a party!"

"I guess so!" He and I stood there for a moment, as I waited for him to say something. Finally he broke the silence.

"You didn't have to get up so early. I would have let you sleep in for a while. You must be tired."

"I didn't want to miss anything! I'm OK. I feel good."

"Good, good."

"What are we going to do today, Merle?"

"Well, it's going to be another big day."

"How so?"

"Well, first of all, I have a very exciting day planned for you."

"I won't even try to guess."

Then Merle dropped the bombshell that I had been dreading, probably more so than I had realized. "Also," he said, "today is my last day."

"What do you mean, your last day?"

"A ship is preparing to leave today. I have to be on that ship. I'm going back home."

That was one of those moments like you might see in a movie, or a TV show, where somebody speaks a devastating line, and the last words reverberate and echo, repeatedly, while the room spins around the recipient of the devastation, who is staggered by the horrible news. I couldn't really say anything in response, and Merle put his hand on my shoulder. "It's time, Ken. You're ready."

Well, I sure didn't feel very ready at that moment. "So you're leaving right now, Merle?"

"No, I'm not leaving right now. We still have a lot scheduled for the day. Let's just enjoy this day, together, and we'll say our goodbyes when it's time."

I don't even remember leaving the house and getting into Merle's car, to tell you the truth. I was trying to appear calm on the outside, but inside I was reeling. I do remember that Merle played some soothing classical music in the car, something by Felix Mendelssohn, I believe. So he must have known that I was upset.

"Where are we going, Merle?"

"First I wanted to stop by the park one last time."

"OK." Frankly, I was glad for anything that might extend Merle's stay a little more, although deep down I knew that the schedule was not going to waver.

We drove to the park, and, like the previous Friday, it was early. We parked the car in the parking lot, this time, and got out. "Let's take a little walk," said Merle. "I just want to enjoy a little nature before I leave."

The finality of that statement hit me like a punch in the gut. We got out of the car and strolled out into the park area. Nobody was even on the basketball court yet. "I don't see any cicadas," I said.

"No, me neither," Merle said. "I'm glad we took the time to really enjoy the cicada from last Friday. I'll never forget that. I can't say that we have any comparable type bug back on Akeethera."

"Do you even *have* bugs on Akeethera?" Suddenly I experienced a sharp pang of regret, as I realized that there were so many questions I had for Merle that I would probably never have a chance to ask him!

"Sure, we have bugs. They're somewhat different than your insects, as would be expected, but they fill the same type of ecological niches. Our planets are quite similar, really, in terms of landforms, and that sort of thing, so the same sorts of creatures have evolved, to a large degree. Swimmers that are like fish, flyers that are like birds, bug-like creatures that are very similar to your insects, furred animals very much like your mammals, reptile-like creatures, and even people that are very much like your people. We're just a few thousand years farther along."

"*Just* a few thousand years!"

"Oh, that's not so much, really. You and I still get along very well, don't we?"

Merle's question bubbled right on down into my gut. A weird, queasy feeling of powerful emotion welled up within me, and my answer was a little wobbly when it came out. "Heck yeah, Merle. You and I get along just great." It hit me that, in a span of a little over a week, Merle had become one of the best friends I would ever have, in my life-- as good a friend as I could ever hope to find. And he was about to leave, forever.

"Oh, look!" Merle said. He stooped down to the ground, and got down on his hands and knees in the grass. "*Gryllus veletis!*"

I didn't remember the scientific name, but when I bent down closer to take a look, I did recognize the same type of cricket which we had seen the previous Friday. "It's the Spring Field Cricket!"

"Very good, Ken! I'm impressed! We just may make an entomologist out of you, yet!"

"I wonder if it's the same cricket we saw the other day."

"No, this has to be a different one. We're quite some distance away from where we were on Friday. A cricket lives his entire existence in a very small area. That is his universe." Merle stood up and took a step back, so he wouldn't accidentally step on the little guy. "Well, I was able to see one last little friend here. That was good." He looked down towards where the cricket was, now well hidden somewhere in the dense grass. "It's been nice to meet you, little friend. I won't forget you, or your brothers and sisters. I'll be thinking of you, from across the galaxy, 97 light years away."

He looked over to me, and I knew he was about to say that he had to leave. "I'm scared, Merle," I said.

"Scared? Why?"

"I don't think I can do this."

"*I* believe in you."

"I don't know *why* you do. Nobody is going to want to listen to me."

"All you need to do is find some open-minded individuals, who have ears to hear. If they truly want to understand the universe, they will easily see that the hyper- dimensional universe makes much more sense than the limited universe they had previously known. If they're stubborn, and just want to hold on to their old beliefs no matter what, there's really nothing that you can say that will change their minds. Those are people that don't want to unlearn what they've learned, period, no matter how 'intelligent' they may appear to be."

"Nobody is going to believe me about *you*, either."

"You can feel free to mention me, or *not* mention me, when I leave-- whatever you think is best! I won't be offended if you feel that you have to say that you came up with the idea of the hyper-dimensional universe all by yourself, just so people don't write you off as some kind of UFO nut or something. But if you think that mentioning me is the best way to go, that is fine, also."

"It's the truth, Merle. I mean, *you* are the truth. I mean-- you know what I'm trying to say. It's true that you told me all this stuff. I wouldn't feel right if I didn't acknowledge what you've done. You've spent years of your life, and traveled a long way."

"Whatever you think is best, Ken."

I didn't know what else to say, at that point, so I just reached out and gave Merle the biggest hug I've probably ever given anybody in my life. Merle hugged me back, and I could see out of the corner of my eye that he was smiling at me.

"It's going to be OK, Ken. You're going to be just fine," Merle said.

"I'm glad you're a lot more confident than I am, Merle."

Merle grinned at me. "I truly am confident, Ken."

"Well, if you're so confident, how about some words of advice? I'm going to need all the advice I can get."

"That's a great attitude to have, Ken. One can always continue learning and growing, do you know what I mean? Just make sure to consider the source of the advice you may receive. In my case, I think you'll be fine." Merle smiled at his own self-promotion, and, putting his hands on his hips, he leaned back a bit and tilted his head up to watch the sky. He stayed in that position for several moments, thinking. Then he looked back over towards me, with a look of resolve on his face. I assumed he was going to give me some specific pointers about how I might get people to believe me about Hyper-Dimensional Relativity, but he surprised me with a much more generalized approach.

"All you need to do, Ken, is focus on being respectful, kind, and giving, and you'll be fine. Embrace the community of your fellow beings, and take the stewardship of your planet into your heart. Understand that I'm not telling you to let people take advantage of you, or to cause harm, but I am saying that people, in general, deserve the benefit of the doubt until they prove otherwise.

"Keep in mind, however, that the intolerant often try to take advantage of the tolerance of others, in order to advance their agendas of hatred or division. In that sense, even tolerance has to have its limits."

I'm sure Merle could see by my facial expression that I was surprised by this train of advice, but he just kept the theme going.

"As far as your own personal life, do not force an unwanted love, but do accept a true, honorable love gladly, with an open heart. Happiness will come to you in profusion, if you simply do that.

"All people have their faults, and moments of weakness. Even knowing that, the act of forgiveness might be the most difficult thing of all to master. But if you can find forgiveness in your heart, that may help keep your happiness from dissolving into anger and resentment.

"If you have a job to do, give it a strong, honest effort. Put good food in your body, make sure to get some physical and mental exercise, and always continue to educate yourself."

I was beginning to wonder if Merle was just getting started, or what.

He continued with barely a pause between sentences.

"Also, take a moment here and there to smell the roses, as they say. Come out here, to the park, and enjoy the natural goodness of your planet. If you can't make it out to the park, or somewhere like this, you should be able to find nature wherever you might be, outdoors-- even indoors, in some cases. Even if you're in the middle of a huge parking lot, or something, there is always the sky to enjoy." Merle reached his arms out straight from his sides, and raised them upwards, looking up to the sky as he did so. "Everything will fall into place, if you follow those principles."

I thought about what Merle was saying, and I couldn't help but feel very doubtful that everything would go as smoothly as Merle seemed to predict. "But our world is so messed up, Merle. There's so much fighting and distrust here—it's not like Akeethera. A lot of people think it's all leading to some kind of an apocalypse, even. People just won't want to listen to me."

Merle smiled at me. "Ken, Ken, Ken," he said. "First of all, do you know what the definition of an apocalypse even is?"

"It's some sort of terrible war, where the world is destroyed, or something like that."

"Not really, Ken. An apocalypse is really a great revelation, about the divine purposes of existence, and about good conquering evil, basically. In practical terms, in our corporeal existence, that would begin with making sure that people are safe from attack, with adequate food, clothing and shelter, and with clean air to breathe, and clean water to drink. Understanding the hyper-dimensional universe can help make those things very attainable, once your science learns to use that knowledge as means for benevolent ends.

"In time, also, your economic theory will become more sophisticated and less zero- sum oriented. After all, they didn't build the pyramids from the top down, did they?" Merle smiled. I'm not exactly clear on what he meant by that last part, but he seemed to enjoy the analogy. I suddenly realized that Merle probably knew exactly how the pyramids were built, but before I could ask him about that, he began talking again.

"What people really want and need are the opportunities to learn and work together cooperatively to produce, as a society, something greater, stronger, and more stable than the whole of its parts. Those are really the starting points, in terms of respecting the fellow inhabitants of your own planet. Love, happiness, and all those other things follow much more easily once people feel safe and have a full belly, dry clothing, a warm house, a sated thirst, and a constructive purpose. Deny any of those, and conflict will arise all too easily!

"Your world has already seen plenty of evil, Ken. Plenty of war, famine, death… you name it, you've had it, and then some. You've already seen more than enough of the very worst human atrocities, often driven by greed and the abominable drive to conquer. Whether in a state-to-state sense, or even in an interpersonal sense, that drive to conquer always tears us apart, and destroys our greater purpose.

"Akeethera went through the same thing, thousands of years ago, but after eons of struggle, we finally turned the tide in favor of peace and reconciliation. *That* was *our* apocalypse, after all the destruction that had tormented our world for so long. Now, Earth may also be getting ready for the redemptive part of the equation. You will also have your transcendent revelations here, on Earth—maybe even starting right here, in Rundle Heights. But those revelations have to come from within, not from without.

"Enjoy your fellow people, Ken, as well as your universe, and know that I will always be out there, supporting you and cheering for you. And not just me, either. You have more friends than you will ever know, out there, who are all on your side, and on Earth's side. I'm not necessarily talking about direct physical intervention, but life in the universe is so precious and difficult to come by, that every living person is cherished by legions, quite literally. Believe me when I say that you are never alone. No human, and no person of any kind, on any world, is ever alone in this universe, no matter your situation or circumstances. That's a simple fact of the universe, and of the Infinite Crystal. That's something that anybody, and everybody, can take strength from. You're not just a tiny fish in the infinite pond, Ken. You're a part of the pond itself, and as such, you are connected with *all* of it."

I followed Merle's gaze upwards, and we both watched a monarch butterfly fly past us, maybe twenty feet up. Far above the butterfly, several cottony white clouds floated peacefully, with a crystalline, azure sky as their backdrop. Far above that, and obscured by the warming rays of our own star--our sun--were trillions of other stars.

Planets twirled around many of those stars, and Akeethera was one of those planets. Countless other civilizations flourished around many other stars, as well, extending throughout our universe, and beyond that, throughout the Infinite Crystal- into the eternally rolling expanse of space/time. I suddenly felt dizzy, but before I lost my balance entirely, Merle reached out and supported me.

"Whoa! Thanks, Merle! I got a little woozy there!"

"You're very welcome, Ken."

"Hey Merle?"

"Yes Ken?"

"Do you think that we-- I mean, what do you think the odds are-- I mean, do you think it's possible that we are being observed, right now, by somebody we can't even see? By somebody we don't even *know* about?"

Merle put his arm around me and smiled. "I wouldn't be surprised, Ken. I wouldn't be very surprised, in the least. And I wouldn't be surprised if somebody *else*, in turn, was observing *them*. And so on! It's probably as true for them as it is for us-- you just can't sneak past that landlady in the kitchen." Then, in classic Merle style, he gave me a big wink.

# CHAPTER
# 41

We walked back to Merle's car, in the parking lot, and got in. I assumed he was going to drive me home, and then leave, but he had saved his biggest surprise for last.

"Hey, Ken," Merle said casually, as if the idea had just occurred to him, "would you like to visit the base, and meet some of my friends? Maybe play a little basketball?"

"You mean on the big black triangle?"

"Yes. What do you think?"

"Are you kidding? Heck yes, I would, totally! I was hoping I could get up there, but I thought I probably wasn't allowed."

"Well, in general, we don't receive visitors from Earth. Truthfully, though, you won't be the first person from Earth to visit the ship. You'll be the second, I believe."

"Who else has been up there?"

"That, I'm afraid, is truly top secret. And it happened several years back. So everybody will be very much looking forward to seeing you and meeting you."

"Well, let's do it, then!" I was very enthusiastic. Eight days prior, I would have been terrified.

"Hold on," said Merle. "I would be remiss, not to mention a few things before we embark."

"OK."

"First of all, please know that people from a large variety of civilizations throughout the galaxy are on that ship. And a few of them are even visiting from other galaxies. About half of the people on the ship are from Akeethera, but amongst the balance of people, you'll see a large variety of different body shapes, different skin colors- all kinds of physical differences, really. Don't expect too much in the way of tentacles, or antennae, or that sort of silly science-fiction thing, though. The evolution of intelligent life on all worlds tends to arise from similar ecological niches, so you tend to end up with very similar models, such as two arms, two legs, two eyes, and a decent number of fingers and toes. However, there *are* beings in the universe, and on the ship, as well, that stretch those boundaries a bit. Also, there are others of very ancient lineages that have evolved beyond our sort of basic physical model, but the few you will see on the ship take a more standard appearance when they are among us."

"Are you talking about the time-savers, who have evolved beyond our basic physical model?"

Merle grinned at that comment. "You really are starting to catch on, aren't you, Ken? The time-savers aren't the only ones in that category on the ship, however. Remember, Ken, it's not unlike your situation here on Earth. There are many, many different types of people on this planet-- tall people, short people, thin people, heavy people, people with glasses, people of different races or nationalities, people with a wide variety of disabilities, people who hold different beliefs about their world and universe, people with different skin colors, people who wear their hair differently, people who dress differently, people who eat different types of food. But all people have much more in common with each other than any of these minor differences. All people need to love, and be loved. All people need shelter, food, fresh water, and safety. All people are trying to build a good life for themselves and for their loved ones. All people hope to be treated with kindness and respect."

"People are people, in other words," I said.

"That's right, Ken. People are people. And it's no different up on the ship. Brothers and sisters are we all."

"And we are all but ducks on the lake."

That comment seemed to please Merle very much. "That is so true, Ken. That is so very true." He smiled at me and chuckled. "You know, Ken, all these people on the ship have volunteered to be a part of this mission. Even so, a long excursion like this can be very difficult on people. Loneliness and deep feelings of isolation can set in, even with so many others on the ship. Homesickness can be a real problem, especially when there's no easy way to return any sooner than originally planned.

"Also, there's a lot of stress involved with returning home after a long superluminal space journey like this. Almost everybody on ship will miss an entire generation of time, back on their home planet. When I return to Akeethera, over 25 years will have passed, while only six years has passed for me, personally. My mother and all of my friends back at the academy will be 25 years older, while I'll be the same age as people who were only young children when I left. All these things can prey on the mind of the space traveler."

"Wow. That sounds really hard."

"It's not so bad, I suppose, if you're trained for it. Still, it can be very hard on people, and sometimes people are looking for something that takes their mind off their own situation. Maybe that's part of the reason people are so excited about your upcoming visit. This will be the event of the trip, and everybody is tremendously looking forward to it."

"Looking forward to me visiting the ship?"

"Oh, yes. Everybody is very much looking forward to your visit."

"Why is that?"

"Well, for one thing, you greatly endeared yourself to everyone by your rescue of Magu. Losing a crew member is the worst possible thing that can happen on a mission like this, and you prevented that disaster from happening on Magu's mission, which has made you a sort of folk hero for many of these people. Secondly, there was a little thing with the ship's computer that really got everybody on your side."

"The ship's computer? What do you mean?"

"Well, everybody was very pleased that I chose you as my third contact. You wouldn't believe how many people told me that they hoped you would get this chance, both before and during the mission. But the ship's computer wasn't so confident. It ran a simulation, and rated you the least likely of the nine total contacts to succeed. By far."

"By far? How far?"

"Well, the computer provided odds on the other eight candidates, ranking their chances of success. Some of them were ranked very high. All of the other eight candidates were either physics professors or physics professionals. In your case, though, the computer didn't even produce a set of odds. It just said that due to your incomplete educational status, your success was "unlikely.""

"Unlikely?"

"That was it. Just 'unlikely'. The computer didn't even put specific odds on it. It actually caused a big uproar, when word got out."

"People were mad that you wasted a pick on me?"

"No, no. People were angry with the computer, if that makes any sense. A lot of people thought it was actually a good thing that you were not a professional, already set in your ways of thinking. Also, your rescue of Magu was a sign of character that the computer didn't seem to be taking into account. So, you see, you really were the "people's choice", as they say here on Earth, from the beginning. That's why everybody is especially happy that you are the last one standing, as they say."

"I guess it is true," I said. "Everybody loves an underdog."

"Well," said Merle, "I never considered you an underdog. I've always considered you to be a very strong candidate. I laughed at that computer report."

We sat there for a moment, and looked at each other. Merle took one last glance around the park, and there was an almost indiscernibly brief moment of darkness. In the next instant, we were high above Earth, and floating maybe 25 yards from the same giant black triangle I had seen out in the forest preserve, just one week prior. An enormous, matte black, metallic panel loomed before us, and as we watched, a tiny rectangular opening appeared, with light shining through from the inside. The opening quickly grew in size until it was perhaps 50 feet across and 20 feet high.

"Merle, this is incredible! Just how big *is* this ship, anyhow?"

"Well, in terms of area, just looking down on it from above, it's roughly 100,000 square feet. But a lot of the ship is subdivided into as many as eight floors, so there's probably more than 600,000 square feet of floor space. It's well over ten million cubic feet in volume."

All I could do was whistle at that. The numbers sounded big. I was too busy trying to peer inside through the expanded opening, to say anything verbally.

"Here we go," said Merle, and we swiftly entered the ship through the opening. In another moment or two, we seemed to have landed on the floor. Looking through the windshield, it was difficult to tell exactly where we were. The walls and floors all shimmered with a warm, uniform light, and it was tough to discern exactly where the wall ended and the floor began. It was almost like when they put the reverse time dilation on me in my kitchen, and everything seemed to be floating in the blue haze. Except here there was no blue haze- just an overall warm, glowing light.

"Don't open the door yet," said Merle.

He really didn't have to say that. I sure wasn't about to just jump right out. I looked behind us and noticed that the opening had closed behind us. I could see that we were inside of a room, or bay. Although it was well illuminated, no light fixtures of any kind were apparent. There was a sort of panel, or screen, on the one side wall, perhaps two feet high and three feet wide, with a much smaller one on the opposite wall. The door which had closed behind us was completely indistinguishable within the back wall, which looked exactly the same as the side walls, except for there being no panel or screen. In fact, I think the entire back wall opened for entry or, presumably, exit. As I looked around, I noticed that the wall in front of us began to open, in the same manner as the wall of the ship had opened. In moments, it was wide open, and we began to move forward.

"OK," said Merle, "we're moving."

I could see that we were leaving the room and entering a horizontally oriented oval tube, or tunnel, that looked much the same as the bay, as far as the lighting in the walls and floor. It was almost hard to tell when we had left the room and entered the tube.

Once it was clear that we were in the tube, though, it seemed like we just stopped moving. "Why did we stop, Merle?"

"We didn't stop! We're actually moving very fast! It's just that, inside this tube, it's almost impossible to sense the motion, and it's gravitationally balanced, too, so you don't feel it internally." Merle smiled at me, and moments later we heard a short clicking sound. "OK, now we're stopped. I just wanted to show you one thing, along the way."

Merle touched the screen, and the doors of our ship opened, with a slick little set of silver stairs- I think there were three stairs—reaching down to the floor beneath the open door of the ship. We were now inside of another room, much like the first room.

"Merle! Isn't it dangerous for me to be out loose in the big triangle ship? Don't I need to be vaccinated, or something?"

Merle smiled at that. "You already have been, Ken, as has every single person on this ship. I didn't think you'd mind, so we did it the first day you rode in the car. Or ship." Merle smiled again and stepped out onto the top stair. With some lingering trepidation, I did the same.

It was odd to step out, and see that our little ship no longer appeared in its camouflaged form, as an automobile, from the outside. I saw it as it really was from the outside- a sleek, gleaming, silver metallic disc-shaped structure, maybe 40 feet across and 15 feet high, with no obvious bolts, welds, or windows anywhere. The ship's doors closed so seamlessly behind us that you would never know doors were ever there, let alone stairs!

The room we were in was fully walled on all four sides. The back wall had closed behind us, much like the outer wall of the ship had closed behind us when we first entered. Now there were panels on three of the four walls, and Merle walked up to the panel on the wall in front of us. When he got there, he turned back to me. "Come on," Merle said, and waved his hand to indicate I should join him. Then he reached his hand up towards the panel, without actually touching it. Suddenly, the wall next to the panel began to open, and an oversized "door" appeared.

Merle gave me another wave and another "come on", and I followed him through the opening. After we passed through the portal, or whatever you'd like to call it, we turned right, and strode into a large, cavernous room, like a large factory-sized area. The lighting was very different in this room, and the first thing I noticed was the amazing profusion of plant life there. There were plants everywhere! I noticed that there were many tall cones, or tapered cylinders which were narrower at the top and wider at the bottom. Most were maybe 40 yards in diameter at the bottom, and 15 yards in diameter at the very top, with some larger and some smaller.

These tapered cylinders had an open, almost antique-looking wrought-iron appearance. They were just about bursting with plants, and they rose up nearly all the way to the ceiling, which must have been at least 100 feet above us. I noticed that the ceiling appeared to be one giant skylight, through which the sun-lit sky was clearly visible. The sunlight inside the room we were in seemed very bright.

"What are those cylinders, Merle?"

"Those? Those are mostly packed with plants for food- what you might call fruits and vegetables, and that sort of thing. Inside each cylinder, there is central access to the plants, for maintenance and harvesting purposes. On Akeethera, just about all of our farming is done inside similar structures-- but most are much larger, and much taller."

"How tall?"

"Oh, some are probably two or three times as tall as the tallest buildings on Earth."

Once again, I let out a low whistle of amazement. "Two or three times! Wow!" Then I noticed that I could actually see figures walking around inside one of the cylinders. "Wow, there are people up there! Are they harvesting?"

"They are probably harvesting some, and also tending to the plants."

Looking around further, I was amazed at the entire set-up. It was very much like a jungle in there. Aside from the cylinders, plants covered much of the floor, or ground, and the walls, too. They were hanging from the ceiling, in long trailing vines, and I could hear the sound of trickling water. There appeared to be quite a few trees, off in the distance, and in open area to our right I could see a group of three figures—children, I guessed, from their diminutive size—playing with some sort of flying, hovering toy. Just as I was about to ask about that, I thought I saw a bird or something shoot past. It was pretty large, and pretty fast. I hadn't gotten much of a good look. "What was that, Merle?"

"That was a flying creature," said Merle, and he obviously didn't want to get into any more details. "Come. We have to leave now. I just thought you might like to see this. There are probably two dozen areas like this, on the ship."

Merle was right about me liking to see that. It was a fantastic place, and I hated to leave, but I had learned that Merle's schedules were not for breaking. We re-entered our ship, and continued on our way through the tube. Before long, we stopped again, in another room, and repeated the disembarking process. This time, after Merle opened the portal, we turned left, and began to walk down another hallway. This hallway seemed to be only for foot traffic.

Merle turned to me as we walked. "You know, Ken, everybody on the ship has been looking forward to this moment, for the last several years. Today is a day of great joy and celebration here. I hope you're ready for some basketball, and a big party!"

Even with Merle's speech about "people are people", and all that, I can't say that I was exactly calm at this moment. My heart felt like it was about to leap right out of my chest. Merle lightly touched me on the right elbow to get my attention. "Come on, Ken. Everybody's waiting."

The hallway was about 15 feet across, and the ceiling had to be about 10 feet high. As we walked along, we passed by dozens of people who were standing in the hallway, in small groups. They smiled and waved at us as we passed by, like they were watching a parade. I almost felt like I should be tossing candy or something to them.

Without exception, they all seemed extremely happy to see us. These people, as Merle had "warned" me, came in all shapes, sizes and colors, and were wearing a great variety of different clothing. Several said "Hello Ken!" in English as we passed by, and I smiled and said "Hello!" back. I was glad for all the friendly attention, but at the same time, I knew it was time to ask Merle a difficult question.

"Merle?"

"Yes, Ken?"

"With all these people, and all these ships up here, why don't they help us more?"

"What do you mean, help you more?"

"I mean, why don't they just come to Earth and stop people from fighting each other?

Why don't they just come down and put an end to our wars? Why don't they feed the hungry, and stop the bad guys?"

Merle stopped walking, and looked deeply at me. "That is a great question, Ken, with many answers. First of all, we are all scientists and explorers, not security forces or fighters. Akeethera, for example, hasn't had any sort of army or military force at all, for over 2000 of your Earth years. We have no need."

"Wow. OK."

"Secondly, in virtually all of your conflicts, both parties are at least partially to blame.

It's likely that each side would say that we were favoring the other side, which would contradict much of what we believe in.

"Thirdly, as far as feeding the hungry goes, making a habit of supplying people with food would lead to an unhealthy dependence on-- shall we say-- manna from the skies."

"I can see that, I guess."

"Fourthly, we inevitably would be attacked, due to the strong feelings of xenophobia still rampant on this planet. Even with the best of intentions on our part, our actions would be perceived as aggression, and we would be left no choice but to defend ourselves. Inevitably, people of Earth would be injured or killed, which completely contradicts our basic mission statement, which is to observe and document, while causing absolutely no harm. We are all peaceable people on this ship."

"What about Atropha? She's fierce!"

"It's true that Atropha can be quite fierce, but her lethality is more bluster than reality.

The girls have been attacked on your planet many, many times, and still, Atropha has never harmed anyone to any significant degree. She *has* melted a few pole-axes, though!

"Also, the girls are not really members of this ship, or our federation, either. They operate independently, and they have their own ship. In truth, the girls, including Atropha, are extremely peaceable people, who nonetheless have chosen to have an in depth, interactive experience on a world that is very violent, compared to their own.

"Our federation conducts most of its business up here, off planet, so we aren't putting our own scientists and researchers in harm's way, which would violate another basic principle of ours, which is to bring everybody back safely. For all these reasons, the guiding principle of the great majority of civilizations monitoring your planet is one of complete non-interference. That is not to say that we may not occasionally lend a hand here or there, in extreme circumstances. Certainly we have lent a hand, many times, throughout Earth's history. But generally speaking, we can only observe and report. Oftentimes that is extremely painful for our researchers, but there is no other way. That is the most difficult part of our missions, without question. All of our own planets went through similar stages of violence, war, upheaval and starvation in the earlier periods of our development, but it is an extremely difficult and painful thing to observe, first hand."

"That all makes sense, Merle. I guess I understand."

"Remember, we are hoping that understanding the hyper-dimensional universe will be a great help to the people of Earth. My mission here was unprecedented, and not lightly undertaken, but we believed it to be worth the tremendous risk. I agreed to try my best to help, in spite of the obvious dangers."

"That's true. You are trying to help, very much, I know."

"I'm not out there zapping the bad guys, but I really am trying to help, and I am most definitely on your side. As we all are."

We continued walking, and I began to hear some kind of loud music playing, nearby. We rounded a bend, and then I saw, at the end of the hallway perhaps 40 feet in front of us, a single person standing. For a moment I admit I was stunned, as she unquestionably resembled a giant praying mantis, in terms of general head and facial appearance. Even so, I somehow divined that she was female, and we continued to walk towards her, without breaking stride. She stood about nine feet tall, with surprisingly bright yellow skin. She was dressed all in green, and as we approached, she held up two blue t-shirts towards us. She held out her arms in what I can only describe as a mantis-like manner. She handed the shirts to Ken, and he thanked her in a language that I couldn't understand. He followed that up with a heartfelt hug, which she returned with what I easily recognizable as a big smile. She then smiled directly at me and held out her hand, which I accepted for a handshake of friendship. Considering that her face and her head looked rather bug-like in general shape, her hand was actually quite pleasing to the touch. I felt really good, actually, as we shook hands, and I could sense her genuine kindness. For a moment I considered offering a hug, myself, but before I got the chance, she said something else to Merle, and he seemed to acknowledge or thank her, quite gratefully. Then, she put her hand up in front of a panel on the wall. Another portal quickly slid open, and she passed through into the hallway on the other side. When the door opened, I could clearly hear that the music was coming from that side. The door swiftly closed behind our tall friend.

"Who was that, Merle?"

"A very important person."

"What did she say to you?"

"Well, she told me that we had made a great choice, in you. She told me that when you barely flinched, and freely extended your hand in friendship, that she knew you were going to have a very good chance."

"A good chance of what?"

"Of success, of course," said Merle.

I'm still not sure I understand, exactly, what Merle had meant by all that. But there wasn't time for any more questions, just then.

"Are you ready?" Merle asked. "Sure, why not?"

"Here, then. Put this shirt on. You can keep it, afterward."

Merle handed me the blue shirt. It wasn't a t-shirt after all-- it was a basketball jersey, and I noticed that it was a common American brand. It looked like something you could buy at any sporting goods store, except this jersey had a large number "10" on the back, underneath the name "Sylvanewski". I couldn't believe what I was seeing. I asked Merle where they got the jersey, and he told me the name of the sporting goods store.

"These were actually purchased in Rundle Heights," Merle said. "We had the letters and the name put on at the store."

I noticed that Merle's jersey featured a number "12", with the name "Akeetheran" on the back. "What are we going to do, Merle, play a game or something?"

"You are catching on faster and faster, aren't you, Ken?" he asked me. "Come on, let's go in."

# CHAPTER 42

Ken put his hand up to the panel, the portal opened, and we strode through a short hallway, with our blue jerseys on. The music was loud, and very foreign and electronic- sounding, with a powerful, pulsating beat. The hallway led to a larger opening, and I could see that we were approaching what appeared to be the inside of a gymnasium, with a regulation-sized basketball court in the center. There was stadium seating surrounding the court on three sides—standard Earth-style stadium-seating-- and they were jam-packed with people, from what I could see, even in the balcony. With a ceiling almost as high as the ceiling in the agricultural area, it seemed like we were entering a mini professional basketball stadium.

There were dozens of people, again in all varieties imaginable, in the hallway leading to the gymnasium. This time, instead of waving at us, the people began cheering and clapping, as we passed by. I exchanged high fives with several of them, who each seemed to know the routine. I noticed that a couple of the hands felt quite different than my own hand. I recall that one hand felt rather leathery, like an alligator purse, perhaps, and another hand felt quite rubbery. We continued on and entered the gymnasium, and I could more clearly see the scope of the throng that had turned out for the exhibition.

There were easily a thousand people or more in the stands, with maybe a couple hundred more standing around the other open areas. The crowd erupted in a giant roar, as we walked in.

Merle and I stood there for a moment, appraising the scene. My heart felt like it was about to beat its way right out of my chest. I was getting quite a cardio work-out, and I hadn't even broken a sweat, yet.

As the roar of the crowd began to dissipate into a hubbub, and then attentive silence, Merle suddenly raised his right hand upwards and waved it vigorously, in greeting. As soon as he began to wave, a great cheer erupted from the assembled throng. It was an enormous, reverberating sound, of whistles and clapping and stomping and frenzied cheering. Hands were waving, and many in the crowd were jumping up and down.

Above the crowd, suspended in the air in a holographic-type display, was a large, rhythmically flashing sign, perhaps 50 feet long and eight feet tall, that read, in English, "WELCOME KEN". Sparkling, very colorful embellishments dashed all around it for additional flair, and additional lighting effects flashed all around the perimeters of the gymnasium area. The cheers and claps thundered through the facility for at least a full minute, as Merle and I stood before them. I am not kidding you when I say that there were even vuvuzelas (plastic stadium trumpets) in that crowd- at least a half-dozen that I noticed. Eventually the cheers and horns began to subside, and I thought that maybe I should do as the Romans do, when in Rome. I raised my hand in greeting and waved, as Merle had done, which resulted in an even louder and wilder sequence of cheering, blaring horns, and flashing lights, which must have carried on for well over two minutes, before things settled down.

Merle turned to look at me. "Well, that was a nice greeting, wouldn't you say?"

Like nearly every interaction I ever had with Merle, everything happened so very fast that day. When the cheering finally settled down, a line of perhaps fifteen resplendent dignitaries came up to us for introductions and greetings. To my astonishment, about ten of the greeters spoke to me in more than passable English, while the other five communicated well enough with bows, smiles and handshakes, with Merle acting as interpreter. As I had observed in the crowd as a whole, the dignitaries represented a cornucopia of various shapes, sizes, skin tones, and clothing. Several, including the captain of the base ship, were from Akeethera, and they shared the bright orange or yellow hair and pebbly skin that are the hallmarks of people from there. I couldn't help but notice that evolution throughout the galaxy did not reliably result in five fingers per hand.

I was surprised when Merle introduced one of the dignitaries by mentioning that he had traveled over 24,000 light years to be a part of the mission. I could tell by Merle's deference to him that he was another important individual in some way. Merle told me it was like traveling all the way from the center of the galaxy, distance-wise. I told Merle I thought that was an amazing distance to travel, and Merle told me that there were a few people on-ship who were from the Andromeda Galaxy, which is about 2-1/2 million light years distant. "You should see their ship," Merle told me. "It's the most incredible ship

I've ever seen, and I've seen a few." I'm still not sure what Merle meant by that, exactly.

The next person in line was memorable for the most amazing colorful patterns of short hair that covered most of his or her body, and for the fact that he or she had traveled across 1700 physical reference frames of space/time, just for a very brief visit, according to Merle. That was one of the dignitaries who seemed to have a bit of a language gap, but nonetheless he or she seemed extremely happy to meet me.

When the greetings of the dignitaries were through, Merle turned back to me and asked how it was going. I told him that I was starting to feel more at home. How could I not, with great reception I had just received?

"Would you like to shoot around a little bit?" he asked me.

"Sure," I said. "Wow, this is such a large gymnasium, Merle! You could hold concerts here."

Merle seemed to love that idea. "That's a great idea! I'm going to pass along your suggestion!" As we strode onto the court, the crowd went wild again, with a repeat of flashing lights and signs, and loud, pulsating music.

I was shocked to see three more players in blue jerseys, already warming up on the court. There were also five players in green jerseys, warming up on the opposite side of the court. I immediately recognized one of the players in blue jerseys. "Latsis!" I shouted. I don't know how she heard me, over the roar of the crowd, but she turned and ran to me, very swiftly, giving me a big hug when she got there. "You're not wearing your hoodie!" was all I could think of to say.

"And you're not on Earth anymore!" she said. "No, I guess I'm not, am I?"

"So are you ready for some basketball?" Latsis asked, and she handed me a basketball.

"We're playing those guys?" I asked, referring to the team in the green jerseys. One of the players in green appeared to be another nine-foot tall female with bright yellow skin. As similar as they looked, I could tell she was not the same person who gave us the shirts. This player could dunk and block shots easily, but she had a hard time catching the ball. There were another two players—I never could tell if they were male or female—who were squat to the ground, and maybe not even five feet tall. They had short legs, long feet, and extremely long arms and hands. These guys were incredible with their ball handling-- even Curly Neal from the Globetrotters would have been amazed, I'm quite sure—but they were agonizingly slow afoot. That didn't matter too much, since it seemed like they could make baskets from half-court or farther, almost without fail. The other two players, apparently a male and a female, were close to me in height. They were stunningly fast, and could catch the ball, but they were not good shooters. The green team, unquestionably, had the most unusual collection of players I had ever seen on one basketball team, and they proved to be evenly matched to our side.

At this point, my other two teammates came over, and Merle was there to introduce us. It was his two mission-mates, from Akeethera!

I think we all got a little emotional out there, meeting for the first time like that. Being friends with Merle, as I was, I felt like I had known them all my life. It was like the roaring crowd didn't exist, as we exchanged handshakes and embraces. I wanted to talk some more with them, since they both spoke a little English, but it was game time.

The game was very different than the last game I played with Merle. This game consisted of two 15 minute halves. A large holographic-type timeclock kept track of the time, and from time to time we took a short break, and they stopped the clock. There was no scoreboard, and no referee. We didn't even keep track of the score. I think it was roughly a tie, but I wasn't keeping track, either. It was the most fun I probably ever had in my life, to this day. The crowd was on their feet the whole time, cheering and blasting their horns, and before the first half ended my stomach was already sore from laughing at the ball-handling exhibition by the two squat players in green. I tried and tried to get the ball from them, but it was quite impossible. I half-expected them to pull out the old bucket of confetti routine on me, as I futilely chased after the ball.

Every time I got a rebound, or someone passed me the ball, the crowd went bonkers. The loud attention took me off my game somewhat, but I didn't mind. That may have been the first time in my adult life when I felt like I was playing a sporting match simply for the pure fun of it, with no conflicting impulses of needing to win getting in the way. I greatly enjoyed it when anybody, on either team, made a nice play.

Undeniably, Merle again was the best player out there, although, frankly, Latsis was close behind him. She told me, afterwards, that she used to come to the triangle base quite frequently just to play basketball with Merle, while Clotro and Atropha were otherwise engaged with other time-saver business elsewhere.

Merle's two mission-mates were OK at basketball, but I was actually better at it than they were. I guess they hadn't been practicing on the court as much as Merle had been. Merle did mention to me, afterwards, that their Eastern mission specialist had gotten very good at table tennis, while the Southern mission specialist enjoyed soccer. "You should see how accurately she can kick a soccer ball," I remember him saying. "I once saw her out here, on this court, kick the ball and hit the backboard, seven times in a row, from beyond half-court, trying to make a basket with a soccer ball. On the seventh try, she banked it into the basket."

# CHAPTER
# 43

After the game ended, both teams got together at center court. We shook hands, and all together, we acknowledged the cheering crowd. Once things started to settle down, Merle shuffled Latsis and me off the court rather quickly. He said that we had to keep moving, and he ushered us into another room, just behind the basket at one end of the court.

It was a mid-sized room, with five chairs, similar to the chairs on Merle's little ship, and a large screen of some kind on one wall. Moments after we entered the room, a panel opened in the opposite wall, and in strode two familiar figures. "Clotro! Atropha!" I shouted. I had certainly come a long way in my feelings towards the three time-savers, and I was so glad to be in the same room with Merle, Clotro, Latsis and Atropha.

"It's great to have the band back together, isn't it?" Merle asked.

Atropha came up to me and gave me a big hug. Just a few days prior, I would have been terrified to see her come at me like that. Now, it simply felt like the natural thing to do. It felt wonderful, in fact. Next came big hugs from Clotro and Latsis, also amazingly wonderful, and we took a few minutes to talk and reminisce about "old times". I caught a little teasing about my budding relationship with Kim, and of course they asked how

L.C. was doing. Then Merle gently interrupted the conversation, and I knew it was time for the time-savers to leave.

Saying good-bye to the three girls was very hard, and I had a final round of heartfelt hugs, as I got to thank each one of them, individually, for all they had done. I was particularly touched when Latsis whispered in my ear, "remember I said that I would help you."

"You sure did," I said. "And I'll never forget it."

And with that, they began to simply fade away, each with a blissful smile on her face. The last thing I heard, as they completely dissolved out of view, was Atropha speaking. Compared with the shrieking roar she once eviscerated me with, this time her voice was a calming purr. "Good-bye, Ken. We're glad that you're the one that was chosen.

You're going to be just fine."

That was the last I ever saw of the time-savers. I still think about them, though, just about every day. I'm holding out hope that they are circling the Earth, even now, biding their time before they stop by some day to pay me another visit. For me, it might be another fifteen years; for them, maybe a few hours, or a week or two at most, probably.

Now, just Merle and I remained in the room. He was standing over by the large screen on the wall, and he motioned me over. "I have some people who would like to say hello, Ken. This transmission is on a bit of a delay, though, so it won't be a two-way conversation. When they're finished, we'll have our chance to respond."

He put his hand near the screen, and a still image came up. There was a lady standing there, with the bright orange hair of an Akeetheran, but without the usual pebbly skin. She appeared to be in her 80s or so, and wearing a shimmering golden outfit, with emerald green accents. Even with the hair and the otherworldly clothing, she looked almost human, really. Behind her was a lush assemblage of exotic, leafy, twining plants in a variety of colors- primarily green, blue and orange. On the wall, behind a large central opening in the arrangement of plants, was a portrait of another Akeetheran, a male. He looked to be in his 20s, also with bright orange hair. He wore a shimmering black and silver outfit, and he had the pebbly skin texture typical of the Akeetherans I had seen on the ship.

"Who is this, Merle? And whose picture is on the wall?"

"The picture is me, just after I graduated from the academy. And the lady is my mom."

"Your mom!? She looks almost human!"

Merle smiled at my "almost human" comment. "Well, let me tell you a quick little story about that. You see, I'm not the first member of my family to have visited this planet."

"What?" This was astonishing news to me.

"It's true. In fact, I was *named* after one of my ancestors who visited this planet about 1500 years ago. The time-savers actually knew him, and spent some time with him, back then, which is amazing to me!

"Much more recently, my father had a mission on this planet, in the summer of 1952, on the West Coast. One day, near the end of his mission, he was walking alongside a river path when he happened to be the only witness to a young woman who accidentally tumbled off the edge of the path, and into the river below. When my father ran over, she was down below, screaming for help in the water. Then she started to sink down under the surface of the water. My father jumped into the river to help.

"Just as he was about to reach her, about 20 feet below the surface, my father was momentarily caught up in a large tangle of discarded fishing line and aquatic plants that he hadn't noticed in the murky water. By the time he got himself free, and was able to reach the woman, who was also tangled up, they were both at the end of their oxygen reserves. With my dad being so far under water, his base ship had a harder time than usual getting a lock on him. Plus, he was holding onto this woman, which complicated things even further, especially considering that she was as close to death, or closer, than my father. So the base ship did what they had to do, and transmitted both of them, together, back to the ship."

"Wow!"

"Wow, indeed. But things got even more interesting after that. My father and this woman spent several days in the Medical area together, recuperating, before she was ready to go back home. But there were a couple of problems."

"Problems? What kind of problems?"

"The first problem was, the woman felt no great compulsion to return. As it turns out, she had lost both parents when she was in her teens, and now she had very little family left- just a single aunt that had moved to Montana many years prior. It was hard for a young single woman, with no family, back in the early 50s. And she was very pleased with the quality of life aboard the ship."

"And what was the other problem?"

"The other problem was, while they were recuperating together in the Medical area, this woman and my father fell in love."

"What?"

"They fell in love. My dad loved this woman, and she loved him and the sacrifice he had made for her. She wanted to stay on the ship, but there was another problem."

"Which was…"

"The other problem was that the ship was about to leave, in a few days, back to Akeethera. And their regulations forbid them from bringing back a native of this planet."

"So what happened?"

"What happened was, my father found a loophole. The day before the ship departed, my father married this woman in an Akeetheran marriage ceremony. She took a new name, "Hannah", for her new life on Akeethera. Now that they were husband and wife, she was allowed to stay with the ship, and come back to Akeethera, regardless of her previous status."

"What was her original Earth name?"

"That, I cannot tell you. I've probably already told you too much," said Merle. "So this lady in the transmission is-?"

"My mother, Hannah."

Well, you could have pushed me over with a feather, at that point. "Merle, how can you be the child of an Earth mother and an Akeetheran father? How can that work, biologically?"

"I'm not their biological child. Biologically, I am the son of my father's sister. But, when I was about three weeks old, about two months after my father returned to Akeethera with his new wife, Hannah, I was orphaned in a very tragic accident at the spaceport, where my parents both worked. My Akeetheran father and Earth mother could not have children, biologically, and they agreed to adopt me and raise me as their own. So they are the only parents I have ever known, other than my first few weeks of life."

"Wow, Merle! Unbelievable!"

"I know. So I'm literally the only person on Akeethera who has a mother from Earth. I suppose that's why I've always been so very interested in this planet. Truthfully, that probably did help me get a spot on this mission, as well."

"So where is your father now?"

"He's still working for the academy. Right now he is traveling, on his final mission before he retires, as a matter of fact, and he wasn't able to get in on this transmission. I will see him in a few months, though. Speaking of the transmission, we'd better get on with it."

One thing that I deeply regret is that I never did get the name of Merle's father. Oh, I have many regrets of questions left unasked, believe me. Although time is endless in the universe, there just never seems to be enough of it, when you get right down to it.

Merle was gazing at his mother's image on the screen. "I can't believe how much she's aged, since I've been gone. It's been quite a few years, back home, compared to my own time frame. I've actually been afraid my mom wouldn't be there anymore, when I got back home. She encouraged me to do this, but I've sure missed her."

We stood there and stared at the image for close to a minute, before Merle snapped out of his reverie. "OK, Ken. Are you ready for the transmission?"

"Sure."

"Here we go, then." Merle put his hand up near the screen again, and Merle's mother began to speak to us.

One thing that I noticed, when Merle's mother began speaking, was her pronunciation of Merle's name. The way she said it was more like "Mur-ell", than "Merle", quite definitely. That's OK, though. I'm still going to stick with "Merle", myself. Merle will always be Merle Akeetheran, to me.

"Hello, Merle," Hannah said, in English. The smile on her face reflected both relief and concern. "I hope you've had a great time on Earth. It looks so beautiful in all the pictures and videos you've sent! Just like I remembered! I envy you, but I don't think I could have gone through all that you did to get the commission... We're so much looking forward to having you back here on Akeethera, and we've heard that your mission has been a great success. Many congratulations, to you and all your crewmates. We're very proud of all of you, and your many years of very hard work.

We're all awaiting your return with great anticipation. I missed you so much, and I love you very much, Merle."

I looked over to Merle, and saw him silently mouth his response, "I love you, too." He looked very happy, but then again he looked like he might be on the verge of tears, as well.

Merle's mother continued. "And Ken, I've been looking forward to finally getting to thank you, personally, for watching over Merle and helping to keep him safe. Merle has told me how very fond he is of you personally, and all the people there. I was very concerned, with him traveling all that way, and spending so much time on a planet he is not totally familiar with. Then, after he was attacked, I couldn't even sleep, for quite a few days. But I feel so much better, knowing that he is with you. As for your mission of spreading the word about the hyper-dimensional universe, I'll be cheering you on all the way, as will everybody on Akeethera, and many more people, throughout the galaxy. Friends forever, Ken." I turned to Merle and flashed him a big smile. "*Me*, keeping *you* safe?" Merle pointed back to the screen. Merle's mother continued, "I have another person, here, who'd like to say hello. You've already met him before, Ken, several years ago."

As soon as she said that, she stepped aside, and a male alien came into the picture, also with bright orange hair and pebbly skin. I immediately knew who it was by simple deduction, if not some hints of his appearance. "Magu!"

"Hello, Ken. Hello, Merle," Magu said. "Ken, I'd like to take the opportunity to thank you, again, for saving my life. I owe everything that I now enjoy to your selfless quick thinking. I will never, ever forget what you did for me, Ken, and how you risked your own life to do so. I wish you full success on your mission, and I want you to know that we are all behind you, 100 percent." He held his hand up towards the screen. "Friends forever, Ken."

Magu said something to Hannah, who was off screen, that I couldn't quite hear, and then he turned back towards the screen and continued talking. "As for you, Merle, my friend, I am looking forward to reminiscing together, about our experiences on Earth. Be safe on your long journey, and we will soon see each other again, in person."

With that, the transmission ended. I looked at Merle, and he looked at me. There is no doubt that it was a powerfully emotional message for us both, and we took a few moments to gather ourselves. Then, Merle put his hand up to the screen again. "Ready to respond?" he asked me.

"Sure."

The screen turned from dark to a light blue, and Merle began to speak. "Mother, it is so very gratifying to see you, and to know that we will be together again soon. My experience down here has been more incredible than I could have ever dreamed of, and I've enjoyed this planet completely, as you had predicted. Still, it'll be so very great to see you again, and to be home again. I've missed you, also, and I love you, also, very much.

"As for Magu, my friend, the groundwork you laid for my mission has proven to be very fruitful. Your suggestion of choosing Ken as our contact was the one single most important contribution that anybody made to our mission. There were plenty of doubters, but you were proven to be quite prescient in your judgement. I'm looking forward to sharing our adventures on this planet with you, when I return. We'll enjoy many great conversations, maybe over a few glasses of fluzle."

At that, Merle turned to me. "Now I will give Ken, here, a chance to say hello as well."

With no time to gather my thoughts, I guess I just spoke from my heart. "Hello, Hannah... I'm very happy, and honored, to meet you. Your son... is a great person. He has worked so hard, and traveled so far, to help my planet, and I've enjoyed my time with him, more than I've ever enjoyed anything. He's opened my eyes to so many things about the universe, and my own planet, which I had never even considered before. I'm going to do my best to make him proud, and to complete the mission after he has left.

"And Magu, it's so great to see you again. I still think about that night, with the car, and I wonder what would have happened had one of us been five feet farther ahead, or behind, on the sidewalk. I'd given up hope of ever seeing you again, and it's so amazing to meet you under these circumstances. This whole experience has been life-changing for me. I feel like I will be completing the mission you started... I will never forget you, Magu."

At that, Merle began to put his hand up towards the screen, but I waved him off so that I could add one final comment. "Hannah, Magu... friends forever."

And with that, Merle put his hand up near the screen, and the blue screen again went black. Merle turned back to me. "Well, Ken, they have me set up on a very fast ship, with a direct route back home, cleared for some very fast hyper-dimensional travel. I will be home before my mother ages much more than another two months or so."

"That sounds great, Merle," I said. But I was thinking it sounded horrible, as far as I was concerned. I was still having difficulty with the thought of Merle going back home.

# CHAPTER 44

"Our time is drawing near, Ken. Do you have any final thoughts or questions?"

"There is one question I've wanted to ask. Well, there are a thousand questions, but one in particular."

"And what is that?"

"Well, what exactly is going on with quantum entanglement?"

Merle was greatly amused with my comment. "Ha! Quite a final question there, Ken! You do get your money's worth, don't you?"

Quantum entanglement was a mysterious phenomenon of quantum mechanics, where fundamental particles, and even larger objects, appear to maintain a physical relationship with each other, in terms of spin direction, for example, even when separated by distances- even substantial distances on the macrocosmic scale. It was a big mystery of physics, and I had hoped I would be able to ask about Merle about it.

"Do you have time to talk about it, Merle?"

"I can give you a brief summary, Ken."

"OK, sure."

"Quantum entanglement typically involves something that is spinning at relativistic velocities- like a spinning photon. It can also work with very rapidly vibrating or oscillating objects."

"OK."

"So, in terms of interaction with the space/time continuum, what is the main difference between a non-spinning object, and an object like a photon, that spins or vibrates or oscillates very fast?"

I saw where he was going with this. "Well, Merle, the photon can interact with any incoming portion of the space/time continuum, from just about zero velocity to nearly $2c$ (twice the speed of light). A non-spinning object, though, is limited to only interacting with space/time at exactly $c$, and nothing else."

"That's right. The photon is a trans-dimensional object, whereas a person, for example, is limited to a single dimensional frame of space/time."

"OK."

"So, for example, let's imagine that you and I manage to entangle the spin of two objects which are spinning at relativistic velocities—in other words, they both are trans- dimensional objects. If we change the plane of spinning of one of the objects, the plane of spin on the other object has to change, in response."

"OK."

"While they are here, in front of us, you and I interact with the space/time continuum only at one velocity- $c$."

"OK."

"But our two objects are spinning at relativistic velocities. So they can interact with the continuum at the full range of velocities that you have described."

"Just like the analogy of the bicycle wheel."

"Exactly, just like that. Now, if we allow one of our spinning objects to leave, at nearly the ratio of space to time, we would be observing both objects in terms of our own slice of space/time, at $c$."

"Right."

"Relative to us, the object that is still in front of us interacts with space/time at the full range of velocities from 0 to $2c$, as you described."

"OK."

"But the second object, now traveling at nearly $c$, interacts with space/time at $c$ to nearly $3c$, relative to us."

"That makes sense. It's shifted almost a full dimension, basically."

"Exactly. So the two objects now share a range of the continuum from $c$ to $2c$, relative to us. That is where their respective physical frames of reference overlap. It's only half the range than what it was when they were motionless to each other, but it is still quite a large range, in comparison to how *we* experience the space/time continuum. That means they still have a very powerful connection."

"OK, that makes sense."

"*We* see the two objects only as they exist at our slice of the continuum, which is at $c$, relative to us. From our perspective, they are getting farther and farther apart, so it seems mysterious that if we change the spin of our remaining object, it will still affect the spin of the second object. They seem to be much too far apart, now, for that to happen."

"Yes, that's what makes it so impossible. Or so *strange*, I guess. Somehow I guess it *is* possible."

"Of course it is possible. These two objects do not just interact with each other at our own slice of space/time, which is $c$. They interact with each other all the way up to $2c$, relative to us, which we don't even perceive."

"Oh... OK."

"Now do you get it, Ken?"

"No, not really."

"How would we, here, perceive that reference frame, at nearly $2c$, if we could, somehow?"

"I'm not sure."

"Well, the second object is nearly keeping up with our expanding frame of space/time reference. So what does that mean in terms of length contraction?"

"It would have a length of just above zero, from our perspective."

"That's correct! So what would that mean in terms of distance traveled, in that dimension? If the length of the object itself is just above zero, in that dimension, relative to us, then how far has it actually traveled in that dimension, relative to us?"

"Not much more than zero?"

"There you have it, Ken! In that dimension, which we cannot even perceive, our two objects are still just about right next to each other. That's why we can still affect the spin of the second object by altering the spin of the first object! That's why quantum entanglement requires that the objects be spinning, or even oscillating, in some way, and why it can act over what appears to us as large distances, in our own physical dimension. In one dimension—our dimension—they are far apart. But in another dimension, they have never really significantly drifted apart. They are still quite close together, in that other dimension. This trans-dimensional linkage is why the connections are so powerful between many of the fundamental particles- they are connected over an entire dimensional range, which is a much stronger connection than one involving a single dimensional frame, as we are used to in our experience."

I was sort of stunned, again, by this idea, and I didn't say anything as I stood there, thinking about it. I was thinking of how close I was to never even asking that question, and therefore not getting the answer. Afterwards, though, I realized that I should have been able to figure the answer out on my own, based on what I had already learned. In retrospect, there were probably a hundred other questions I dearly wish I had asked, instead, for my final question, but what's done is done. I am very grateful for anything and everything that I did learn from Merle while he was here. Luckily, he offered up one final tidbit, unsolicited.

"So, Ken, do you have a new appreciation for the mass/energy equivalence equation now?"

"The mass/energy equivalence equation?"

"Yes, *e* equals mc squared."

"What about it?" I was a little confused, as usual. Merle smiled at me. "It's all the same," he said.

"What do you mean?"

"Well, energy is interchangeable with mass, in the equation, is it not?"

"Yes. Mass is just energy in another form, really."

"And what about space/time?"

"What about it?"

"It's the other part of that equation, Ken-- the '$c$' part; the ratio of space to time in the universe. Space/time itself is a type of energy, Ken, isn't it? It's energy that interacts with matter in the form of a wave, at $c$. It causes change—entropy, really-- via the passage of time. So, essentially, time is just energy of a different stripe. Time is non-quantum energy, whereas mass-energy is quantum energy."

I nodded my head at Merle, as he spoke. I didn't want to miss anything, and I had a strong feeling that time was getting tight now.

"So you see, Ken, within a matrix of relativistic three-dimensional space, everything in the universe is energy, really." He then moved closer, within inches of my face, and looked me right in the eyes before delivering the kicker, in a loud whisper. "Whatever *energy* is!" And with that, Merle broke out into a most tremendous grin, along with his now very-classic "case closed" wink. He turned towards the screen, chuckling out loud just a bit.

Merle waved his hand in front of the screen again, and it suddenly became a window to the outside world. Far below us was the Earth, in all its living glory. I walked over by the screen to look more closely.

"There it is, Ken. Glorious, isn't it?"

"Heck yes. Majestic." I was looking at my own portion of the planet, this time, including most of North, Central and South America.

"Yes, it is majestic. My memories of this planet will fill flash upon my inward eye for the rest of my days."

Merle turned to face me more directly. "Your planet has been circling this star, and spinning on its axis, for almost five billion years now, and it will continue to do so for a very long time to come, Ken. I just want you to remember that your life involves much more than this mission, and the physics that we have discussed. Remember the other important things that we discussed in the park this morning. If you continue to do all those things, you will have greatly succeeded, in my eyes.

"As far as the physics goes, if you struggle to find anyone with ears to listen, and they are too stubborn to broaden their minds with new ideas, and you find nothing but a brick wall in your path, the world will continue spinning on its axis, and orbiting the sun. Life will go on, and hopefully your planet will still find its share of successes and positive outcomes. Perhaps, someday, someone else will come along who may have better success in elucidating the idea of a hyper-dimensional universe to a wide audience. You cannot force people to open their minds, or to unlearn what it is they have already learned. All you can do is try your best to persuade them, Ken.

"To me, from what I have seen, you are already well on your way to a successful mission. If you can somehow enlighten your world to the realities of the hyper- dimensional universe, that will be the cherry on top, as they say. But if you are unable to, don't let that destroy your own happiness, or your ability to be a respectful person. Those things are much more important than understanding how the universe works, when you get right down to it. Those are things that you can do, by yourself, and you can always find pleasure and satisfaction in your life, if you do those things."

Merle stopped talking, so I took my eyes off Earth, and looked back towards Merle. "OK, Merle. I'll try to do my best, and I'll try not to be too upset if I can't get anybody to listen about the physics. But I will try my hardest to get them to listen, somehow. "

"Good, good, Ken. That's a great approach. And now, I have a parting gift for you. You can keep the basketball jersey, as I said, but I have a little something else for you. Something I'd love for you to have." Merle walked over to one of the chairs in the room, and reached behind it. He pulled out a silvery bag, similar to a shopping bag, and reached into it. He pulled out an old fashioned vinyl record album. I saw the familiar prism on the black background on the front cover, and I couldn't let him do it.

"Merle, no! You can't give me your copy of 'Dark Side of the Moon'! That is your most prized possession! You told me that yourself!"

Merle gave me that beatific smile of his. "Don't you see, Ken? That's why I want you to have it. Besides, I don't have a record player, anyhow."

"You think I have one? I don't!"

"That's OK, Ken. It belongs with you, not me. Besides, I have something *else* to remember you by." He reached into the same bag and pulled out the set of blueprints that Professor Jonmur had given him. "You have a copy of these, and I have a copy, too. *This* is now my most beloved possession, Ken." He reached out, physically opened up my left hand, and placed the "Dark Side of the Moon" album in my hand. "Please, Ken. This will make me so happy. I will cherish the thought that you have the album, now. You'll have the only copy of 'Dark Side of the Moon' on Earth that has traversed over 250 light years through space."

I literally didn't know what to say. I was choked up with emotion at the magnitude of the gift, and I have to admit some tears were welling up within me.

We were now face to face to each other, and Merle put his right hand upon my left shoulder. "Friends forever, Ken."

I did the same, placing my right hand upon Merle's left shoulder, so that we were connected together, each of our right arms on the other's left shoulder. "Friends forever, Merle."

Then came the now familiar brief darkness, and suddenly I was standing in my kitchen, holding the album in my hand. I dropped the album on the floor, and fell down to my knees on the tiles, as my legs involuntarily gave out from under me. Thankfully, I had the presence of mind to adjust my position a bit, so I didn't land on the album with my knees. I knelt there, for a good minute or more, holding my head in my hands, until L.C. came over, and rubbed up against me. I regained my composure, and put the album—fortunately undamaged—on the counter. Then I took L.C. up into my arms, and went over to the couch. I gave her the best petting she ever had, and she purred, even more loudly than when Atropha had petted her.

# CHAPTER 45

Four years have now passed, on Earth, since Merle went back home to Akeethera, and so much has happened during that time. Sometimes it's hard for me to believe it's only been four years.

Charles and I are just about to have our Fourth Annual Cleanup of Streamside Park. It's really gone much easier than I had expected, especially with so much help from Kim and Mr. Teakurk. Tommy Marnel came through with a lot of volunteers, too, like he had promised me he would, and Harold has been the hardest worker of anybody, especially in terms of dragging debris out of the creek. The first year was sponsored by The Enterprise and the waste disposal company owned by Mr. Teakurk's brother in law, as well as the accounting firm that Kim was interning at, but we have an additional ten sponsors lined up for the Fourth Annual Cleanup, which helps.

We've also made some real progress towards getting some substantial improvements made to the entire setup of the park, per Maximillian Jonmur's original plans. By next summer, we hope to begin new construction, to reverse some of the channelization of the stream that was made so long ago. Charles is very confident that we can get it all done over the next five or six years, which would be incredible. The Park District has even agreed to partially rebuild the southern portion of the park, and to shuffle some of the playing fields back to that side, especially since several of the fields were due for major upgrades, anyhow. The county was able to procure some federal funding for the project also, so we could bring in some of the heavy equipment needed.

Ronny and Eva got married, as I had mentioned earlier, and they have their first child on the way. I'm very happy to announce that they won't be alone in their matrimony, since Kim and I are engaged now. We're going to be married in the spring.

Neddie moved in with Kim's aunt before last winter, and I've been renting the house from Neddie since then. L.C. was included in the deal, as had been promised, and I am very much enjoying having her as a pet again. Neddie still stops by from time to time, more to see L.C. than for any other reason, I think, although she says she's just making sure that everything is fine with the property.

Kim works in Rundle Heights, and I work in the next town over, so after Kim and I exchange vows, we are going to purchase the house from Neddie. We joke around that it's only so L.C. won't have to move. It will be nice living next to Walter, and my parents are really happy about it. It's very sweet of Neddie to hold back on the sale of her house, and to help us carry out our plan, as she has. She is very excited about the wedding, too, and she has even been helping Kim's mom with some things.

As far as the physics goes, Merle was correct in thinking it might be like beating my head against a brick wall. Even when I have managed to get a physicist to at least listen to me, usually they are too busy to give me much more than the time of day. The few physicists that I was able to talk to at any length have all said the same thing: my theory is way too big, and I need to choose only one aspect at a time in order to have a chance at getting it published in a peer-reviewed scientific journal. The problem is that hyper-relativity doesn't fully make sense unless you look at the entire scheme of the thing. If I tried to only publish any one portion of the idea, it would automatically be rejected as being "impossible." If I tried to publish the entire theory at once, it would be automatically rejected as being "too large." The whole thing is an incredibly frustrating Catch-22 situation.

It was Charles who recommended that I choose a different tack, and present the idea of hyper-relativity in the form of a reality-based novel. After giving it much thought, I agreed that it was worth a try.

Everybody seems to want to classify it as "science fiction", though, due to the Merle angle.

I know this for certain: I am not going to give up! I will try, try, try, and try again, for the rest of my life, if necessary. If I never succeed, the Earth will keep on spinning, and orbiting the sun, and life on our planet will go on-- hopefully successfully, as Merle had suggested.

I still frequently visit Streamside Park, to clear my mind and enjoy the natural goodness of our world, as Merle had suggested I do. Sometimes I park on the street, like we did that first day, and sometimes I park in the soon-to-be demolished lot, where a couple of great adventures in Merle's "car" began. It's a special treat if I see a cicada, or a monarch, or a crow, or children playing, or a game out on the basketball courts.

Sometimes I stop by the courts to say hello to Tommy, if he's out there. A lot of the times Harold is out there, too, and we never fail to reminisce about that amazing day when Merle showed us all a thing or two about Earth basketball.

What I *always* look for, though, when I am out at the park in the summer months, are crickets. If you ever see somebody at the park who is kneeling on the sidewalk, or in the grass, and he is sobbing out loud while weeping profusely, then that is probably me, looking down at a Spring Field Cricket.

But those are not tears of pain, or tears of desolation. Those are tears of pure joy, and gratitude. Those are the tears of my Enlightening.

The End

Made in the USA
Lexington, KY
08 May 2019